Learning and Analytics in Intelligent Systems

Volume 26

The main aim of the series is to make available a publication of books in hard copy form and soft copy form on all aspects of learning, analytics and advanced intelligent systems and related technologies. The mentioned disciplines are strongly related and complement one another significantly. Thus, the series encourages cross-fertilization highlighting research and knowledge of common interest. The series allows a unified/integrated approach to themes and topics in these scientific disciplines which will result in significant cross-fertilization and research dissemination. To maximize dissemination of research results and knowledge in these disciplines, the series publishes edited books, monographs, handbooks, textbooks and conference proceedings.

More information about this series at http://www.springer.com/series/16172

Eugenia Politou · Efthimios Alepis · Maria Virvou ·
Constantinos Patsakis

Privacy and Data Protection Challenges in the Distributed Era

Eugenia Politou
Department of Informatics
University of Piraeus
Piraeus, Greece

Efthimios Alepis
Department of Informatics
University of Piraeus
Piraeus, Greece

Maria Virvou
Department of Informatics
University of Piraeus
Piraeus, Greece

Constantinos Patsakis
Department of Informatics
University of Piraeus
Piraeus, Greece

ISSN 2662-3447 ISSN 2662-3455 (electronic)
Learning and Analytics in Intelligent Systems
ISBN 978-3-030-85445-4 ISBN 978-3-030-85443-0 (eBook)
https://doi.org/10.1007/978-3-030-85443-0

This Springer imprint is published by the registered company Springer Nature Switzerland AG
The registered company address is: Gewerbestrasse 11, 6330 Cham, Switzerland

The authors would like to thank their families for their continuous support

Series Editor's Foreword

It has been stated that "*Data in the 21st Century is like Oil in the 18th Century*" [9, 5]. This statement eloquently illustrates the fact that data is acting as a thread that unites digital, biological, physical, and energy sciences and technologies towards the 4th Industrial Revolution [7] and Society 5.0 [4, 2] of human civilization.

Indeed, during the past two decades, humanity has been witnessing an explosion of data collection through the use of a variety of both traditional and novel sensing devices (e.g. wearables). Besides traditional economic figures, all sorts of additional and diverse data are currently collected which, to name just few, include data related to mobility, traffic, GPS, outdoors activity, travel, planning, user habits, shopping, customer behaviour, entertainment, social interactions, education, medicine, biology, health, forensics, or energy consumption [10, 5].

Consequently, the amount of the collected data is huge. To get a feeling of it and of its rate of increase, it suffices to look at some recent statistics and near-future predictions by market intelligence companies. While in 2010 only about 2 Zbytes (i.e. 2 x 10^{12} Gbytes) of data were created worldwide, the total data created, captured, or replicated (but not necessarily stored) in 2018 was estimated to 18 Zbytes, i.e. to 18 x 10^{12} Gbytes [6]. On the other hand, the World Economic Forum estimated that 45 Zbytes of data were generated in 2020 [11]. Furthermore, the projections of the market intelligence company IDC indicate that, by the year 2025, the world's data will grow to ten times the 2018 figure; i.e., the global datasphere will rise to about 180 x 10^{12} Gbytes [6, 8]. As eloquently illustrated by IDC, a data of that size would require a CD stack long enough to circle the Earth 222 times and 1.8 billion years to download at the currently available Internet speeds [6]. By comparison, a good laptop in the year 2021 carries a disc storage capacity of 128–1000 Gbytes.

Good quality data is highly desirable, and the information extracted from them benefits nearly all aspects of society, including economic development, healthcare and caregiving, famine, poverty and inequality reduction, disaster prevention, and efficient energy production and distribution [4, 2, 10]. However, and despite these significant benefits, extensive data collection and processing for information extraction pose dangers to humanity and personal privacy and good care must be taken to manage the societal implications of these new technologies [1].

To safeguard against data misuse, especially highly sensitive data, and to protect personal data and the privacy of individuals, the General Data Protection Regulation (GDPR) of 25 May 2018 established a new, EU-wide regime. Overall, the GDPR was mostly welcomed. However, in its Article 17, the GDPR grants individuals with the right to request their personal data to be erased from all the available sources to which they have been disseminated ("Right to be Forgotten") [3]. Enforcing the provision of Article 17 is highly non-trivial in today's technological state, especially in a mobile, distributed, decentralized, big data-driven, and artificial intelligence-empowered environment. Clearly, enforcing legislation requires new research and technological advancements beyond the state of the art. To assess the size of this challenge, one has to consider that in parallel we have the integration of blockchains in a wide range of applications. One of their unique features that drives their adoption is their immutability. Similarly, we have the use of machine learning and artificial intelligence algorithms which depending on their use may come up with outcomes which were not initially known, not communicated to the data owner, and without her consent, violating this way GDPR from another perspective. The above create a very complex situation where different cutting edge and emerging technologies, such as 5G, IoT, big data, blockchain, machine learning, and artificial, intersect within an highly heterogeneous ecosystem of stakeholders so GDPR enforcement becomes a huge challenge.

To address these technological challenges, Drs. Politou, Virvou, Patsakis, and Alepis have just delivered a high-quality monograph entitled "Privacy and Data Protection Challenges in the Distributed Era". In the monograph, consisting of 11 chapters, the authors:

1. highlight the provisions of the GDPR and its Article 17,
2. present consequences that arise from state-of-the-art technologies such as mobile computing and sensing, blockchains, and the InterPlanetary File System (IPFS),
3. identify challenges and "gaps" on the state-of-the-art enforcement of Article 17, and
4. propose efficient solutions to the previous which mitigate conflicts and, at the same time, safeguard the privacy and rights of individuals.

The book also includes a detailed discussion of the effect of extensive use of digital technologies in the past 18 months due to measures taken to protect public health from the COVID-19 pandemic and additional safety and privacy concerns that it has caused.

Authoring a book to address privacy and data protection challenges in the distributed era is a highly anticipated but certainly not an easy task. The authors, conducting world-class research in this area, are most qualified and well-versed in the corresponding problems, and their own research provides excellent solutions. While reading through the book, I was impressed by both the breadth of topics addressed and the depth of the provided solutions. I found that, as the book is structured, the reader is offered the choice to either go directly to specific chapters in it that are of particular interest to him/her or to study through extensive parts or even the entire monograph and get informed on broader aspects of privacy and data protection in the distributed era. Thus, I expect the book to be useful to both the specialized researcher seeking information on specific sub-areas within privacy and data protection in the distributed era as well as to the newcomer who seeks to get involved in this exciting research field.

In summary, I warmly congratulate Drs. Politou, Virvou, Patsakis, and Alepis for their superb work and I highly and unreservedly recommend the book to professors, graduate students, practitioners, and other specialists in privacy and data protection in the distributed era and to general readers, all of whom I expect to greatly benefit from it in their researches.

July 2021

Prof.-Dr. George A. Tsihrintzis
Department of Informatics
University of Piraeus
Piraeus, Greece

References

1. N. Bostrom, M.M. Cirkovic, in *Global Catastrophic Risks* (Oxford University Press, 2011)
2. Cabinet Office, Government of Japan (2018) Society 5.0. URL https://www8.cao.go.jp/cstp/english/society5_0/index.html
3. European Union, Regulation (EU) 2016/679 of the European Parliament and of the Council of 27 April 2016 on the protection of natural persons with regard to the processing of personal data and on the free movement of such data, and repealing Directive 95/46/EC (General Data Protection Regulation). Official J. Eur. Union L 119 (4 May 2016), pp. 1–88
4. i-SCOOP, From Industry 4.0 to Society 5.0: the big societal transformation plan of Japan (2015). URL https://www.i-scoop.eu/industry-4-0/society-5-0/
5. D. Parkins, The world's most valuable resource is no longer oil, but data (2017). URL https://www.economist.com/leaders/2017/05/06/the-worlds-most-valuable-resource-is-no-longer-oil-but-data
6. D. Rydning, J. Gantz, D. Reinsel, in *The Digitization of the World from Edge to Core*. Tech. rep., IDC (2018)
7. K. Schwab, The fourth industrial revolution. Currency (2017)
8. STATISTA, Volume of data/information created, captured, copied, and consumed worldwide from 2010 to 2025 (2021). URL https://www.statista.com/statistics/871513/worldwide-data-created/
9. J. Toonders, Data is the new oil of the digital economy. URL https://www.wired.com/insights/2014/07/data-new-oil-digital-economy/
10. G.A. Tsihrintzis, D.N. Sotiropoulos, L.C. Jain, in *Machine Learning Paradigms: Advances in Data Analytics* (Springer, 2019)
11. World Economic Forum, Technology futures: Projecting the possible, navigating what's next (2021). URL http://www3.weforum.org/docs/WEF_Technology_Futures_GTGS_2021.pdf

Preface

This book examines the conflicts arising from the implementation of privacy principles enshrined in the *General Data Protection Regulation* (GDPR), and most particularly of the *"Right to be Forgotten"* (RtbF), on emerging computing platforms and decentralized technologies. Essentially, the RtbF allows the possibility for individuals to request the erasure of their personal data from all available sources to which they have been disseminated when certain conditions are met. While the enforcement of the GDPR in 2018 was mostly welcomed, its RtbF provoked widespread skepticism and heated debates due to the drastic consequences when enforced in the era of big data and blockchains.

To elaborate on the impact of the RtbF upon modern information systems, we first identify the various notions of forgetting and the need to be forgotten, including the case of revoking consent, both in the social and technical context. In this vein, we investigate the challenges of implementing the RtbF in organizational and business processes such as in the established backup and archiving procedures. Towards seeking GDPR compliance, we evaluate technical frameworks and architectures in terms of fulfilling the practicalities for integrating the RtbF into current computing infrastructures.

Afterwards, we study two ground-breaking innovations of our times: mobile ubiquitous computing and decentralized file storage systems. Specifically, we delve into the progress of mobile affective computing and the state of the art in decentralized p2p networks, i.e. the blockchain and the *Inter-Planetary File System* (IPFS), and we explore their privacy risks in relation to the GDPR principles. Against this background, we identify and discuss the privacy risks imposed by ubiquitous mobile computing practice and research when combined with big data algorithmic processing to infer sensitive personal details. We particularly study the risks of profiling which are further elaborated in the tax and financial context. Then, we review the emerged trends towards accountable algorithmic processing and we explore strategies to mitigate the risks of aggressive profiling and discriminatory automated decisions. Furthermore, we investigate the extent to which the GDPR provisions establish a protection regime for individuals against those risks and we propose domain-specific countermeasures.

Next, we delve into the impact of the RtbF in emerging decentralized technologies such as the blockchains and IPFS. Our analysis demonstrates that the challenges of enforcing the RtbF on these environments are not trivial. In view of this, we investigate the incompatibility between the blockchain's immutability and the RtbF, and we review advanced cryptographic methods for introducing restricted mutability into blockchain's design. As these methods, however, present certain limitations, off-chain workarounds based on the IPFS are increasingly adopted by many blockchain projects. Considering that storing files in the IPFS network does not remove the burden of erasing them should the RtbF be claimed, we study the extent to which the IPFS protocol complies with the RtbF. Our analysis reveals that the IPFS does not adequately adhere to the RtbF since it is not currently feasible to efficiently enforce data erasure across its entire network.

Our main contribution towards resolving the conflict between the IPFS and the RtbF is the proposal of an anonymous protocol for delegated content erasure that could be integrated into the IPFS to distribute efficiently and securely a content erasure request under the RtbF. The protocol complies with the primary IPFS principle to prevent censoring; therefore, erasure is only allowed to the original content provider or her delegates. A formal definition and security proofs are provided, along with a set of experiments proving the efficacy of the proposed protocol. To the best of our knowledge, this is the first application-agnostic proposal to align the IPFS with the RtbF and to endorse its GDPR compliance. Hence, we firmly believe that our work adds real value to the IPFS in terms of its privacy enhancement and, consequently, contributes significantly to its future adoption by applications that are processing personal data.

We conclude our book by investigating the interplay between privacy and COVID-19 pandemic. In this context, we explore the invasiveness of digital technologies utilized to monitor people's health, such as the mobile contact tracing apps and health immunity passports, and we discuss their impact on individuals' privacy and data protection rights.

Piraeus, Greece
June 2021

Eugenia Politou
Maria Virvou
Constantinos Patsakis
Efthimios Alepis

Contents

Acronyms

AEFI	Automatic Exchange of Financial Information
AEOI	Automatic Exchange of Information
AI	Artificial intelligence
AISP	Account Information Service Provider
API	Application Programming Interface
ASPSP	Account Servicing Payment Service Providers
BMA	Blackboard Manager Agent
CA	Certificate Authority
CJEU	Court of Justice of the European Union
CNIL	Commission Nationale Informatique & Libertes
COVID-19	Corona Virus Disease 2019
CRS	Common Reporting Standard
DACs	Directives on Administrative Cooperation
DAG	Directed Acyclic Graph
DAO	Decentralized Autonomous Organization
DApp	Decentralized Application
DG TAXUD	European Commission Directorate General for Taxation and Customs
DFS	Decentralized File Storage
DHT	Distributed Hash Table
DLT	Distributed Ledger Technology
DPA	Data Protection Authority
DPD	Data Protection Directive
DPoS	Delegated Proof-of-Stake
DRIPA	Data Retention and the Investigatory Powers Act
DSHT	Distributed Sloppy Hash Table
EBA	European Banking Authority
EBP	European Blockchain Partnership
EBSI	European Blockchain Services Infrastructure
ECJ	European Court of Justice
EDPB	European Data Protection Board
EDPS	European Data Protection Supervisor

EnCoRe	Ensuring Consent and Revocation
ERP	Enterprise Resource Planning
EU/EEA	European Union/European Economic Area
FA	Financial Authorities
FATCA	Foreign Account Tax Compliance Act
FPB	Financial Privacy Broker
FFI	Foreign Financial Institution
FI	Financial Institutions
GPA	Global Privacy Assembly
GDPR	General Data Protection Regulation
HCI	Human–computer interaction
HDFS	Hadoop Distributed File System
HMRC	Her Majesty Revenue and Customs
HTTP	HyperText Transfer Protocol
ICO	Information Commissioner's Office
ICT	Information and Communications Technologies
IGA	Intergovernmental Agreement
IoT	Internet of Things
IPFS	Inter-Planetary File System
IPLD	Inter-Planetary Linked Data
IRS	Inland Revenue Service
MCAA	Multilateral Competent Authority Agreement
MiFID2	2nd Markets in Financial Instruments Directive
ML	Machine learning
MPC	Multi-party computation
MS	Member state
NHS	National Health System
OECD	Organization for Economic Co-operation and Development
RtbF	Right to be Forgotten
P2P	Peer-to-peer
PBFT	Practical Byzantine Fault Tolerance
PC	Personal computer
PIN	Personal Identification Number
PISP	Payment Initiation Services Providers
PKI	Public key infrastructure
PoA	Proof of Authority
PoS	Proof of Stake
PoW	Proof of Work
PPT	Probabilistic Polynomial Time
PSD2	2nd Payment Services Directive
RTS	Regulatory Technical Standards
SMC	Secure Multi-party Computation
SMS	Short Message Service
SPV	Simplified Payment Verification
TCP/IP	Transmission Control Protocol/Internet Protocol

TPP	Third-party payment
TTL	Time To Live
UMA	User-Managed Access
UN	United Nations
WP29	Article 29 Working Party
XACML	eXtensible Access Control Markup Language

Chapter 1
Introduction

The rapid pace of technological developments over the last few years reveals more and more new cultural and ethical challenges. The privacy of individuals is one such issue that is heavily questioned under the pressure of technological progress. As a matter of fact, most of the times privacy is put entirely aside in favour of innovations that promote the intensive use of personal information. Unfortunately, this is the case when technology precedes its time: researchers often face dilemmas of following innovative technological approaches while ensuring appropriate levels of privacy or other ethical principles. Similarly, users also have to make decisions concerning the disclosure of their personal information on the basis of a difficult trade-off between privacy and the advantages stemming from data sharing.

While privacy is commonly discussed in the context of data protection, these two terms are not synonymous. In fact, privacy and data protection are two closely interrelated notions, albeit not identical. Although both derive from well established fundamental human rights under the European Law, they actually represent two different values: privacy generally refers to the value of protecting one's private life, whereas data protection refers to the value of limiting or controlling the conditions on the processing of data relating to an identifiable individual. Nevertheless, data protection and privacy overlap to a great extent. In consequence, privacy – besides data protection – is always within the scope of contemporary data protection legislations, even though when this is not explicitly stated such as in the case of the newly enacted legislation across the EU, the *General Data Protection Regulation* (GDPR) [2], that defines the new regime for the protection of personal data of individuals.

The GDPR has been put into effect on the 25th of May 2018 replacing the 1995 European Data Protection Directive (DPD) which up to then constituted the international standard against which all data protection initiatives, in and out of Europe, were judged. While the principles encompassed by the GDPR were mostly welcomed, its enforcement also provoked great scepticisms due to its severe impact on the processing of personal data within and outside the EU territory [3]. Of its provisions, the most radical and controversial one that has been subject to heated debates is Article

E. Politou et al., *Privacy and Data Protection Challenges in the Distributed Era*,
Learning and Analytics in Intelligent Systems 26,
https://doi.org/10.1007/978-3-030-85443-0_1

17 that anticipates the *Right to be Forgotten* (RtbF). In simple terms, the RtbF – along with the provisions for withdrawing consent – allows the possibility of individuals to request the erasure of their personal data from all the available sources to which they have been disseminated when certain conditions are met (Article 17(1)). Although the RtbF has been viewed as an evolution of the already established rights under the EU national data protection laws, according to legal experts the right is actually a novelty as it has a broader scope than any existing right. For instance, a unique feature that differentiates it from the rights granted by the previous legislation is its retro-activity according to which individuals have the right to request the retroactive erasure of all of their personal data.

Undeniably, the legislation of the RtbF raised great concerns due to its collision with many organizational processes and current business practices. It also gave rise to long worldwide debates among scholars within law, privacy and computing disciplines as the implications of imposing the GDPR are tremendous for data controllers processing personal data both within and outside the EU. Beyond any doubt, the consequences of encompassing the GDPR provisions, and particularly the RtbF, in contemporary information systems are immense, whereas its integration into the design of new technological advancements is currently disputable. By all accounts, enforcing the RtbF in the era of big data, Internet of Things (IoT), and blockchains is not trivial. As these modern technological advancements rely on the collection, processing and decentralized storage of a vast amount of personal information, the notions of privacy and data protection are heavily challenged. Consequently, the strict provisions of the GDPR on personal data processing is seen as being incompatible with the progress of big data analytics, machine learning algorithms, ubiquitous computing, blockchains and other decentralized peer-to-peer (p2p) networks such as the *Inter-Planetary File System* (IPFS) [1].

1.1 Book Objectives

Against the aforementioned background, this book aims to investigate the impact of the GDPR, and of the RtbF in particular, on a wide range of contemporary organizational processes, current business practices, and emerged state-of-the-art technologies, as well as to provide specialized solutions for bridging the identified discrepancies. In what follows, we describe our main research objectives which are:

1. To explore in depth the various notions of forgetting and the need to be forgotten, including the case of revoking consent, both in the social and in the technical context. In this regard, we review all controversies around the new stringent definitions of consent revocation and the RtbF in reference to their impact on privacy and personal data protection rights.
2. To elaborate on the impact of the RtbF upon modern information systems and to assess emerged state-of-the-art technologies in terms of fulfilling the technical practicalities for integrating the new forgetting requirements effectively into cur-

rent computing infrastructures. In this respect, and towards seeking GDPR compliance, we evaluate technical methods, architectures, and frameworks – existing either in corporate or academic environments – and we review their weaknesses and strengths in reference to users' full control over their personal data, and particularly their effective erasure from third-party controllers to whom the data have been disseminated.

3. To discuss the GDPR requirements with respect to their impact on the backup and archiving procedures stipulated by the modern security standards. In this context, we examine the implications of erasure requests on current IT backup systems, and we highlight a number of envisaged organizational, business and technical challenges pertained to the widely known backup standards, data retention policies, backup mediums, search services, and ERP (Enterprise Resource Planning) systems.
4. To identify the risks to privacy and data protection rights posed by two advanced technological trends of our times that have been emerged in parallel and independently of the GDPR but they are used increasingly nowadays for storing and processing personal data: the ubiquitous mobile computing and the decentralized p2p networks.
5. To delve into the privacy risks imposed by big data algorithmic processing to infer sensitive personal details such as people's social behaviour or emotions. In this respect, we deal with the threats of predictive algorithmic assessments and profiling methods that make important decisions about individuals and we analyse the specific case of profiling and automated decision-making on the basis of the GDPR. To mitigate these risks, we explore implementation challenges and countermeasures against the discriminating processing of individuals and groups in the context of financial privacy.
6. To investigate the consequences of encompassing the GDPR provisions, and particularly the RtbF, in the design of new technological advancements such as the blockchain. In this regard, we identify major inconsistencies between the blockchain and the RtbF resulting mainly due to the blockchain's by design immutable nature. Towards resolving this contradiction, we comprehensively review the state-of-the-art research approaches, technical workarounds, and advanced cryptographic techniques that have been put forward to enable restricted mutability on blockchain environments – aligning thereby blockchains with the GDPR – and we discuss their potentials, constraints and limitations when applied in the wild to either permissioned or permissionless blockchains.
7. Given that several blockchain projects are adopting the IPFS (a prominent decentralized p2p file sharing and storage network) for storing the actual personal data off-chain in order to comply with the GDPR's RtbF obligation, we first investigate how the IPFS controls the dissemination of personal data within its network, and then we carry out research for providing an efficient solution to align the IPFS with the RtbF. To this end, we work towards formally introducing an anonymous protocol for delegated content erasure that could be easily integrated into the IPFS to distribute a content erasure request securely among all the IPFS nodes.

8. To discuss the security and privacy challenges of COVID-19 digital surveillance technologies. To this end, we introduce the basic building blocks and technical charasteristics of the contact tracing mobile applications and health immunity passports, and we discuss the impact of these digital efforts on data protection and individuals'privacy.

1.2 Book Structure

This book starts in Chap. 1 by introducing the concepts and topics researched and analysed in the rest of the Chapters and by outlining the key areas of our research and contributions. In Chap. 2 the notions of privacy and data protection are presented whereas their corresponding rights are further elaborated in the tax and financial context.

In Chap. 3, we introduce the new regulation for data protection across the EU, the GDPR, and we discuss its main principles. Among others, we thoroughly investigate one of its controversial right that has raised great concerns and led to long debates: the *Right to be Forgotten (RtbF)*. To that end, we elaborate on the various notions of forgetting and the need to be forgotten both in the social and in the technical context. However, as consent revocation can potentially trigger the effective erasure of personal data under the RtbF, the *right to withdraw consent* is also analysed and its various characteristics and implementations are also presented.

Following the analysis of the RtbF in Chaps. 3, 4 explores the challenges of implementing the RtbF on contemporary information systems. In this regard, we delve into the impact the GDPR may have upon enforcing it on already established backups and archiving procedures stipulated by the modern security standards. Towards achieving GDPR compliance, we evaluate existing methods, architectures and frameworks, existing either in business or academic environments, in terms of fulfilling the technical practicalities for the implementation and effective integration of the RtbF into the current computing infrastructures.

Provided that the RtbF is found to collide with various technological developments, in Chap. 5 we present and examine two state-of-the-art technologies emerged over the recent years: the ubiquitous sensing and mobile computing, and the decentralized p2p networks. Delving further into these emerging technologies, we analyse the cases of mobile affective computing, as well as these of the blockchain and the IPFS.

In Chap. 6, we research into the privacy challenges arising from the ubiquitous mobile computing and sensing applications. In this regard, we analyse the privacy risks imposed by big data analytics and machine learning practices, and we thoroughly discuss the risks of profiling which are further elaborated in the tax and financial context. Then, we explore strategies towards mitigating these privacy risks, and we investigate the extent to which the GDPR protects against these threats, especially against aggressive profiling and automated decision-making. Furthermore, we introduce some countermeasures in the context of financial privacy that can adhere to the privacy requirements of the GDPR.

Chapter 7 deals with the privacy challenges imposed by the blockchain technology. Beyond other risks, blockchain immutability – according to which tampering with data already stored in blockchains is nearly impossible – is being examined regarding its compatibility with the RtbF enshrined in the GDPR. Inevitably, as the RtbF is found to be in conflict with the blockchain tampered-proof nature, the controversy over the mutability of blockchain protocols is thoroughly discussed. To this end, current technical methods, practical workarounds, and advanced cryptographic techniques for balancing immutability with the RtbF are comprehensively reviewed in terms of their applicability in public and private blockchains.

In the light of blockchain's incompatibility with the RtbF, in Chap. 8 we explore the properties of the IPFS which is employed widely to store personal data off-chain and, as such, to eliminate the privacy risks of blockchains. As this solution, however, moves the onus of removing the personal content to the IPFS network, the compatibility of the IPFS with the RtbF is investigated. Towards conforming the IPFS to the RtbF requirement, we formalize an anonymous protocol for delegated content erasure that can handle securely any content erasure requests across the entire IPFS network. Formal definition and security proofs of the proposed protocol are also provided, along with a set of experiments that prove the efficacy of the proposed protocol.

The privacy challenges pertained to the COVID-19 pandemic are discussed in Chap. 9 where we investigate the invasiveness of digital technologies utilized to monitor people's health, such as the mobile contact tracing apps and health immunity passports, and we discuss their impact on individuals' privacy and data protection rights.

In Chap. 10 we underline and explore open issues and future research directions pertained to the challenges of the state-of-the-art technologies considered in this book for their effective alignment with the GDPR. Finally, in Chap. 11 we conclude our book by presenting and discussing the main findings of our research.

References

1. IPFS, InterPlanetary File System (2021). https://ipfs.io/
2. European Union (2016) Regulation (EU) 2016/679 of the European Parliament and of the Council of 27 April 2016 on the protection of natural persons with regard to the processing of personal data and on the free movement of such data, and repealing Directive 95/46/EC (General Data Protection Regulation), Official Journal of the European Union, Law 119 (4 May 2016), pp. 1–88
3. E. Politou, E. Alepis, C. Patsakis, Forgetting personal data and revoking consent under the GDPR: Challenges and proposed solutions. J. Cybersecu. **4**(1), tyy001 (2018)

Chapter 2
Privacy and Personal Data Protection

Abstract Privacy nowadays is commonly discussed in the context of data protection. While privacy and data protection are not synonymous, they overlap to a great extent. In consequence, privacy —besides data protection—is always within the scope of contemporary data protection legislations. As both terms derive from well established fundamental human rights under the European Law, they are closely interrelated notions—albeit not identical. They actually represent two different values and, legally speaking, two distinct fundamental rights under the European Law. In this Chapter, we discuss the values of privacy and personal data and we analyse what the respective right to privacy and to personal data protection entails. We also discuss the meaning of privacy in the tax and financial domain taking into account the principles of data protection.

2.1 Introduction

Privacy and personal data protection are two interrelated terms that are often used interchangeably; nonetheless, they actually constitute two discrete and different notions. The idea of privacy in Europe derives from concepts such as human dignity and the rule of law. Modern conceptions of privacy have begun to be developed following the experiences of fascism in World War II and communism in the post-war period. Within the European law, there is a distinction between "privacy" and "data protection" which defines these two concepts as closely related, and often overlapping each other, but not as synonymous [1]. Privacy generally refers to the protection of an individual's "personal space", while data protection refers to limitations or conditions on the processing of data relating to an identifiable individual. Nevertheless, as legal scholars note [2, 3], data protection and privacy overlap on a way whereby data protection is both broader and narrower than privacy. It is narrower because it only deals with the processing of personal data, whereas the scope of privacy is wider. It is broader, however, because it applies to the processing of personal data, even if the latter does not infringe upon privacy.

E. Politou et al., *Privacy and Data Protection Challenges in the Distributed Era*, Learning and Analytics in Intelligent Systems 26,
https://doi.org/10.1007/978-3-030-85443-0_2

In what follows, we discuss the values of privacy and personal data and we analyse what the right to privacy and to personal data protection entails.

2.2 The Value of Personal Data

In general, personal data refer to the information relating to an individual. While many Data Protection Acts define personal data in more or less similar terms [4–6], the GDPR elaborates a little further on their definition (Article 4) [7]:

> "Personal data" means any information relating to an identified or identifiable natural person ("data subject"); an identifiable natural person is one who can be identified, directly or indirectly, in particular by reference to an identifier such as a name, an identification number, location data, an online identifier or to one or more factors specific to the physical, physiological, genetic, mental, economic, cultural or social identity of that natural person;

This definition clarifies that personal data are any information that can be used on its own or with other information to identify, contact, or locate an individual. Although in 2009 the European Commissioner for Consumer Protection Meglena Kuneva compared personal data to oil in order to illustrate how information pertaining to individuals had become a crucial asset in the digital economy,[1] it was Clive Humby who first coined the parallelism of data to oil back in 2006 to denote that data are just like crude: "*It's valuable, but if unrefined it cannot really be used*".[2] The World Economic Forum [8, 9] has also described personal data as a new asset class for which a complex ecosystem of entities collecting, analysing, and trading personal information has emerged. Spiekermann et al. underline in [9] that personal data are seen as a new asset due to their potential for creating added value for companies and consumers by providing services hardly imaginable without it. As Acquisti et al. describe [10], personal information has both private and commercial value and often exploiting its commercial value entails a reduction in privacy and sometimes even in social welfare overall. Stated otherwise, most privacy issues originate from the two different markets, the market for personal information and the market for privacy, which are actually two sides of the same coin [10].

2.3 The Value of Data Privacy

Even though privacy was introduced as a right in 1890 by Warren and Brandeis [11], it was only the last three decades that it has been extensively discussed in its various forms and contexts, mainly due to the blow of computing and informational sciences. As Introna noted back in 1997 [12], privacy has emerged as a philosophical issue in the late 1960 and since then is discussed in great controversy within the academic, legal,

[1] http://europa.eu/rapid/press-release_SPEECH-09-156_en.htm.

[2] http://ana.blogs.com/maestros/2006/11/data_is_the_new.html.

and social circles. Still, no universally accepted definition of privacy exists. Privacy can be seen either as the right to be left alone [11], the "*power to selectively reveal oneself to the world*" [13], or as the control over personal information or even as the freedom from judgement by others [12]. Post explains [14] that "*privacy is a value so complex, so entangled in competing and contradictory dimensions, so engorged with various and distinct meanings, that I sometimes despair whether it can be usefully addressed at all*". He categorizes privacy as two contradicting and antagonizing rights, privacy as dignity and privacy as freedom. Privacy as dignity safeguards the socialized aspects of the self; privacy as freedom safeguards the spontaneous, independent, and uniquely individual aspects of the self. Each seems necessary in a civilized society, but they are also incompatible with each other [14]. According to Solove [15], the term "privacy" is an umbrella term, referring to a wide and disparate group of related things and cannot be understood independently from society since privacy, in its core, is a social artefact, and without the context of society, there would be no need for privacy. He also proposes a taxonomy to privacy threats [15] focusing on the different kinds of activities that impinge upon privacy and demonstrates the different harms and problems.

The value of data privacy, as Acquisti et al. explain in [16], is frequently debatable since what people decide their data worth depends critically on the context in which they are asked and, specifically, how the problem is framed. Therefore, privacy valuations are not normally or uniformly distributed, but U-shaped, clustered around extreme, focal values due to the dramatic gap that has been observed between subjects' "willingness to pay" to protect the privacy of their data as well as subjects' "willingness to accept" money in order to give up privacy protection. Although usually people do concern about their privacy and the adequacy of data protection regulations, they cannot understand the potential threats unless they have a personal stake in it [17]. A survey conducted some years ago among US adults[3] found that the majority of adults feel their privacy is being challenged and showed that people give important weight to the idea that privacy applies to personal rights and information, whereas 91% of adults in the survey agreed that consumers had lost control over how personal information is collected and used by companies. Paradoxically enough, although in earlier times control over personal data may have been best undertaken by preventing the data from being disclosed, in an Internet-enabled society users are showing an increased demand for more data collection, which illustrates that they do not necessarily want more privacy as if concealment, but primarily want more control and transparency on the way their data are being used and reused [18, 19].

2.4 The Rights to Privacy and to Data Protection

Legally speaking, privacy and data protection both represent two distinct fundamental rights under the European Law which defines the first one as a substantive right, thus is created to ensure the protection and promotion of interests of human individuals

[3] http://www.pewinternet.org/2014/11/12/public-privacy-perceptions/.

and society, whereas the second as a procedural, thus operates at the level of setting the rules, methods and conditions through which substantive rights are effectively enforced and protected [20].

The stand-alone fundamental right to data protection was foreseen in the 1981 Convention for the Protection of Individuals with regard to Automatic Processing of Personal Data (Convention 108), and it was recognized as a fundamental, autonomous right under the Article 8 of the Charter of Fundamental Rights of the European Union enacted by Lisbon Treaty in 2009 [21]. It has been pointed out that the principles underpinning the human right to data protection reflect some key values inherent in the European legal order, namely privacy, transparency, autonomy and non-discrimination [22]. Therefore, under an instrumental conception, it can be argued that the right to data protection could serve as a safeguard not only for privacy but also for all fundamental rights [2]. The right to privacy, on the other hand, is also a well-established right by the European Convention on Human Rights (article 8) [23] which entered into force in 1953.

As for the newly enforced regulation for data protection in Europe, the General Data Protection Regulation (GDPR), it is expressly framed in terms of rights, with Article 1 noting that the regulation "*protects fundamental rights and freedoms of natural persons and in particular their right to the protection of personal data*". Despite the fact that there is not any reference to the right to privacy throughout the GDPR text, the concept of privacy is implied in most of its recitals and articles.

Due to their relation to the aggressive profiling techniques that are discussed in Sect. 6.2.4, in what follows we examine the notions of privacy and data protection rights in the tax and financial context.

2.5 Privacy in the Tax and Financial Domain

Historically, tax privacy related to the non-disclosure of tax information has long preoccupied authorities, society, and citizens. Since the late 19th century, where tax returns were considered to be public documents deliberately disclosed in order to increase social compliance [24], until the 21st century, where tax data are combined with other financial and personal information in order to assess the potential tax evaders, tax privacy has stimulated lengthy debates among tax and legal scholars. In the US, where the privacy is regulated through narrow sectoral laws, it was not until 1975 that the Congress established the principle of tax privacy in a statute for the first time [24, 25]. In Europe, however, the rights of tax and financial privacy derive from the fundamental rights of privacy enshrined in the European Convention of Human Rights [23].

Up to recently, tax privacy has been discussed in the extensive literature as synonymous with "tax confidentiality" with references to privacy harms when tax information is disclosed to the public or sent for secondary uses to other agencies [24, 26]. On the other hand, financial privacy is usually taken as the right of individuals to determine what financial information about them should be known to others and most commonly refers to the maintenance of confidentiality of customer information about financial transactions [27].

On the contrary, the right to data protection enshrined in the data protection laws is not simply about the confidentiality of the data being gathered and exchanged but it gives the data subject far more extensive rights, such as the right for personal data to be collected and exchanged only for lawful clearly identified purposes and not to be retained longer than necessary for the identified purposes [28, 29]. Taking into account the principles of data protection, tax and financial privacy nowadays have a broader meaning which includes the adverse privacy implications of the extended collection of information in order to build detailed profiles of people's tax and financial behaviour either for profit or for auditing. The privacy risks arising from such aggressive profiling practices are described in depth in Sect. 6.2.4.

While the amount of the private non-financial information, and especially sensitive, that compose the tax and financial information is extraordinary, Hatfield [26] underscores that the identification of some details as private does not mean necessarily that it is unjust to collect them because privacy concerns do not always outweigh other factors. Instead, they should be weighed against these factors. Indeed, as other scholars explain, the right to tax and financial privacy is not absolute since an absolute right to privacy would make any modern tax system unworkable [24, 27, 30]. Hence, it is uncontroversial that privacy can be compromised to defend other rights or social interests. However, the presumption is that the onus is on authorities to provide a compelling reason why privacy should be compromised, rather than the onus being on citizens to show why privacy should be upheld. In the context of taxation, in particular, individuals only have procedural safeguards (e.g. notification, consultation or intervention) and not a substantive right to privacy. Yet the absence of those procedural rights might constitute an infringement of the substantive right to privacy [30, 31].

References

1. J.F. Abramatic, B. Bellamy, M. Callahan, F. Cate, P. van Eecke, N. van Eijk, E. Guild, P. de Hert, P. Hustinx, C. Kuner et al., *Privacy Bridges: EU and US Privacy Experts In Search of Transatlantic Privacy Solutions* (Technical Report, University of Amsterdam, 2015)
2. R. Gellert, S. Gutwirth, The legal construction of privacy and data protection. Comput. Law Secur. Rev. **29**(5), 522–530 (2013)
3. S. Kulk, F.J. Borgesius Zuiderveen, Privacy, freedom of expression, and the right to be forgotten in Europe, in ed. by J. Polonetsky, O. Tene, E. Selinger, *Cambridge Handbook of Consumer Privacy* (Cambridge University Press, 2017)
4. Data Protection Act, Ireland, *Ireland Data Protection Act 1988 (Data Protection Act 2003, as amended)*. http://www.irishstatutebook.ie/eli/1988/act/25/enacted/en/html (2003)
5. Data Protection Act, UK, *UK Data Protection Act 1998*. http://www.legislation.gov.uk/ukpga/1998/29/contents (1998)
6. Data Protection Directive, Directive 95/46/EC of the European Parliament and of the Council of 24 October 1995 on the protection of individuals with regard to the processing of personal data and on the free movement of such data. Off. J. Eur. Union **L281**, 31–50 (1995)

7. European Union, Regulation (EU) 2016/679 of the European Parliament and of the Council of 27 April 2016 on the protection of natural persons with regard to the processing of personal data and on the free movement of such data, and repealing Directive 95/46/EC (General Data Protection Regulation). Off. J. Eur. Union **L119**, 1–88 (2016)
8. Schwab, K., Marcus, A., Oyola, J., Hoffman, W., Luzi, M.: *Personal Data: The Emergence of a New Asset Class*. http://www3.weforum.org/docs/WEF_ITTC_PersonalDataNewAsset_Report_2011.pdf (2011)
9. S. Spiekermann, A. Acquisti, R. Böhme, K.L. Hui, The challenges of personal data markets and privacy. Electron. Markets **25**(2), 161–167 (2015)
10. A. Acquisti, C. Taylor, L. Wagman, The economics of privacy. J. Econ. Lit. **54**(2), 442–492 (2016)
11. S.D. Warren, L.D. Brandeis, The right to privacy. Harvard Law Rev. **4**(5), 193–220 (1890)
12. L.D. Introna, Privacy and the computer: why we need privacy in the information society. Metaphilosophy **28**(3), 259–275 (1997)
13. E. Hughes, A cypherpunk's manifesto, in *The Electronic Privacy Papers*, (Wiley, 1997), pp. 285–287
14. R.C. Post, Three concepts of privacy. Geo Law J. **89**, 2087 (2000)
15. D.J. Solove, A taxonomy of privacy. Univ. Pennsylvania Law Rev. **154**, 477–564 (2006)
16. A. Acquisti, L.K. John, G. Loewenstein, What is privacy worth? J. Legal Stud. **42**(2), 249–274 (2013)
17. A. Raij, A. Ghosh, S. Kumar, M. Srivastava, Privacy risks emerging from the adoption of innocuous wearable sensors in the mobile environment, in *Proceedings of the SIGCHI Conference on Human Factors in Computing Systems* (ACM, 2011), pp. 11–20
18. J. Ausloos, The "right to be forgotten"—Worth remembering? Comput. Law Secur. Rev. **28**(2), 143–152 (2012)
19. E.A. Whitley, Informational privacy, consent and the "control" of personal data. Inf. Secur. Techn. Rep. **14**(3), 154–159 (2009)
20. N.N.G. de Andrade, Oblivion: the right to be different from oneself: re-proposing the right to be forgotten, in *The Ethics of Memory in a Digital Age* (Springer, 2014), pp. 65–81
21. Charter of Fundamental Rights of the European Union, *Charter of Fundamental Rights of the European Union, 2012/C 326/02* (2012)
22. Y. McDermott, Conceptualising the right to data protection in an era of Big Data. Big Data Soc. **4**(1), 2053951716686994 (2017)
23. European Convention on Human Rights, *Convention for the Protection of Human Rights and Fundamental Freedoms (European Convention on Human Rights, as amended) (ECHR)* (1950)
24. P. Schwartz, The future of tax privacy. Natl. Tax J. 883–900 (2008)
25. P.M. Schwartz, D.J. Solove, Reconciling personal information in the United States and European Union. Calif Law Rev. **102**, 877 (2014)
26. M. Hatfield, Privacy in taxation. Florida State Univ. Law Rev. **44**, 579 (2016)
27. J.C. Sharman, Privacy as roguery: personal financial information in an age of transparency. Publ. Admin. **87**(4), 717–731 (2009)
28. P. Baker, Privacy rights in an age of transparency: a European perspective. Tax Notes Int. del **82**, 583–586 (2016)
29. R. Calo, Privacy and markets: a love story. Notre Dame Law Rev. **91**, 649 (2015)
30. R.S. Avi-Yonah, G. Mazzoni, *Taxation and Human Rights: A Delicate Balance*, vol. 520 (U of Michigan Public Law Research Paper, 2016)
31. P. Baker, Taxation and the European convention on Human Rights. Br. Tax Rev. **40**(8), 298–374 (2000) (International Bureau of Fiscal Documentation)

Chapter 3
The General Data Protection Regulation

Abstract The enforcement of the GDPR on the 25th of May 2018 has caused prolonged controversy due to the severe impact on the processing of personal data under this new regulation. Of its provisions, the most radical and controversial one is the "*Right to be Forgotten*" (RtbF). In simple terms, the RtbF—along with the provisions for withdrawing consent—allows individuals, under certain conditions, to request the retroactive erasure of all of their personal data. In this chapter, we present the main data protection principles enshrined in the GDPR, and we explore the various notions of forgetting and the need to be forgotten—including the case of revoking consent—both in the social and in the technical context. In this regard, we review all controversies around the new stringent definitions of consent revocation and the RtbF in reference to their impact on privacy and data protection rights. Furthermore, we document frequent consent misuses as well as current frameworks for managing and revoking consent. We also shed light on the common misconception that equates the RtbF defined under the GDPR with the one enforced by the CJEU decision at the Google Spain case in 2014.

3.1 Introduction

Upon the enactment of the General Data Protection Regulation (GDPR) on 25 May 2018 across the European Union (EU), new legal requirements for the protection of personal data were enforced for the data controllers operating within the EU territory. While the principles encompassed by the GDPR were mostly welcomed, two of them—namely the *right to withdraw consent* and the *Right to be Forgotten (RtbF)*—caused prolonged controversy among privacy scholars, human rights advocates, and the business world due to their pivotal impact on the way personal data should be handled under these new legal provisions.

In this Chapter, we firstly present the main data protection principles enshrined in the GDPR, and then we review all controversies around the new stringent definitions of consent revocation and the RtbF in reference to their implementation impact on our digital era. On this account, we document frequent consent misuses as well as current frameworks for managing and revoking consent. We also discuss the differences

E. Politou et al., *Privacy and Data Protection Challenges in the Distributed Era*,
Learning and Analytics in Intelligent Systems 26,
https://doi.org/10.1007/978-3-030-85443-0_3

between forgetting and the need to be forgotten, and we shed light on the common misconception about the RtbF defined under the GDPR and the one enforced by the Court of Justice of the European Union (CJEU) decision at the Google Spain case in 2014.

3.2 Introduction to the GDPR

On 27 April 2016, after four years of drafting, lobbying and negotiations among the EU Member States (MSs) and many affected organizations,[1,2,3] the GDPR was agreed and finalized, whereas on 4 May 2016 its final text published in the Official Journal of the EU [1]. Following a two year implementation period, the GDPR applied across the EU on the 25th of May 2018.

The GDPR's introduction aimed at replacing the Data Protection Directive 95/46/EC (DPD) [2] introduced in 1995 which, being a directive, left some room for interpretation during its transposition into individual national laws. In addition, the rapid change in data landscape caused, among others, by the explosion of ubiquitous computing and big data practices, had led to the necessity for another update to the regulatory environment within the EU. Yet, the radical changes brought in by the GDPR affect severely businesses operating within and outside the EU territory. Most importantly, as a regulation and not a directive, it immediately became an enforceable law in all the MSs, and hence, it contributed to the harmonization of the data protection laws across the EU, enhancing at the same time both data protection rights and business opportunities in the digital single market.

The regulation accomplishes its objectives, on the one hand, by strengthening the well-established data protection principles already specified in the DPD, such as the consent and the purpose limitation, and on the other, by encompassing new principles such as the right to data portability, the obligation for data protection impact assessments, and privacy by design, among others. Since its first draft in 2012, much debate has been taken place among scholars and law experts about the fundamental changes that it introduces. Two particular GDPR principles, however, rocked the boat of legal, academic and business world: the reintroduced concept of consent along with its revocation as well as the newly introduced RtbF. They both caused prolonged controversy due to the drastic consequences of enforcing these new requirements in the era of big data, decentralized services, and the IoT.

Hereafter, we first underline the core data protection principles enshrined in the GDPR and then we review all controversies around the new stringent definitions of consent revocation and the RtbF in reference to their implementation impact on digital environments.

---

[1] http://www.eugdpr.org/gdpr-timeline.html.

[2] http://edps.europa.eu/data-protection/data-protection/legislation/history-general-data-protection-regulation_en.

[3] http://www.jimmcguigan.co.uk/EJCNews_July2013DATA%20protection_bothstories.pdf.

3.3 The GDPR Data Protection Principles

According to many legal scholars, the most important contribution to the personal data processing by the GDPR is the choice of the instrument itself: the moderation of the EU data protection landscape through regulation, rather than a directive, constitutes a turning point for the EU signalling a forced exit of this particular field of law from the MSs level to the EU level [3]. Nevertheless, to the disappointment of many privacy advocates, the final version of the GDPR includes a large number of provisions that leave room for national interpretations and approaches depending on the culture, focus and priorities of the supervising authorities.

Taken into account that the DPD constituted the international standard against which all data protection initiatives, in and out of Europe, were judged [3], the main data protection principles enshrined in the GDPR are revised but are broadly similar to the principles set out in the Directive: fairness, lawfulness and transparency (Article 5(1)(a)); purpose limitation (Article 5(1)(b)); data minimization (Article 5(1)(c)); accuracy (Article 5(1)(d)); storage limitation (Article 5(1)(e)), accountability (Article 5(2)), integrity and confidentiality (Article 5(1)(f)). Yet, the GDPR brings the novelty of explicitly imposing organizations to enshrine "*data protection by design and by default*" (Article 25) enforcing measures such as data minimization as a standard approach to data collection and use. Despite the long history of incorporating the privacy by design principles into privacy preserved systems, privacy by design has been formulated into concrete applicable design principles by Cavoukian in 2011 [4, 5]. These principles encapsulate concepts such as data minimization, purpose limitation, transparency and control, all anticipated by the GDPR in the form of "data protection by design and by default" provisions (Article 25). Furthermore, the GDPR extends the provision on automated individual decision-making to include profiling cases as a prime example of enabling individuals to control their personal data in the context of automated decision making (Article 22). In that regard, it acts as a crucial function towards mitigating the risks of big data and automated decision making for individual rights and freedoms [6].

Besides the above, the regulation introduces some new rights for data subjects such as the right to data portability (Article 20) which ensures the interoperability of subject's data and requires data to be provided in a structured, commonly used and machine-readable format and, when required, the controller to transmit the data directly to another controller. While the portability right seems reasonable and has been welcomed by the majority of public and private organizations, there are some other rights that have raised great concerns and stipulated long debates between scholars within law, privacy and computing disciplines. Specifically, the GDPR introduces the right to withdraw consent (Article 7(3)) and the RtbF (Article 17), two secondary rights that derive from fundamental concepts of data protection. These two controversial rights will be extensively discussed hereafter.

3.4 Consent and Revocation

Consent aims at providing legitimate grounds to data controllers for collecting, processing or even disseminating personal data for secondary use. While consenting is one among several available legal grounds to process personal data under all data protection regulations up to date, is assuredly the most global standard of legitimacy and most likely to engender user trust [7]. In that regard, the various notions of consent have been intensively debated for their utilization in online environments and research projects. Even though consent may have various forms with similar flavours (such as informed, explicit, unambiguous or broad), each of these is quite diverse in nature.

Informed consent can be said to have been given based upon a clear appreciation and understanding of the facts, implications, and consequences of an action. Flashing back, informed consent was a cornerstone of the Nuremberg Code ethical guidelines which originated in the pre-World War II Germany, and they specify that informed consent is not only essential for safety, protection and respect for participants, but also for the integrity of research itself [8]. While informed consent was originally a process for obtaining permission in medicine before conducting a healthcare intervention on a person, it has also been adopted in the field of computer science when applications require the collection and processing of data from live subjects. As explained by Reynolds in [9], "*to be informed, consent must be given by persons who are competent to consent, have consented voluntarily, are fully informed about the research, and have comprehended what they have been told*".

Depending on the methodology, the population, the topic under study and the level of risk, informed consent may be implied or explicit, active or passive, and written or oral. For obtaining an explicit consent, participants should give consent through an explicit affirmative action, such as by answering a specific question, in written or oral form, about their willingness to participate. On the other side, broad consent involves agreeing to a broad set of potential secondary future uses under a particular governance framework and it has been widely adopted as the standard practice in many genetic registries and biobanks. Broad consent is also a standard practice for most big data projects where their most innovative secondary uses can't be imagined by the time of data collection.

The academic discussions on whether the user consent in online research and marketing should be informed and explicit [10–13] or broad [14, 15] is heated and the relevant literature is split, while many academics have argued in both ways [16, 17] or in favour of additional countermeasures [18–20]. Meanwhile, other conceptions of consent have been put forward, like the collaborative consent [17, 21], the dynamic consent model [20, 22] which is actually a tool that could better facilitate the process of obtaining any form of consent, and recently, the notion of meta-consent [23].

Another rising tension for the use of consent comes from the potential benefits of big data analysis and the need for explicit or informed consent [24]. Edwards in her notable work [7] examines the issue of obtaining meaningful prior consent in the era of IoT, big data and the cloud, especially when data are collected in public, as in the context of "smart cities". As Barocas and Nissenbaum have been intelligibly expressed [24], "*big data extinguishes what little hope remains for the notice and choice regime*" since upfront notice is not possible in case the value of personal information is not apparent at the time of collection when consent is normally given. Let alone when new classes of goods and services usually reside in future and unanticipated uses [25–27]. This motivated many radical voices to argue against the need for consent which may jeopardize innovation and beneficial societal advances [28] and therefore, its role should be circumscribed with respect to prospective data uses and, in specific cases, consent should not be required to legitimize data use. Still, for other scholars [3, 27] consent requirements are the last defence for individuals against the loss of control on their personal information processing and thus, eliminating or reducing the need for informed consent cannot be accepted uncritically and seemingly without public debate, particularly if democratic ideals are valued.

The notion of consent revocation, or withdrawal, has also been brought into light recently, with many arguing for a right to revoke consent and for a more user-friendly and personalized consent mechanism [29, 30]. Indeed, when individuals are given the opportunity to grant consent to the use of their personal information as a primary mean for exercising their autonomy and to protect their privacy, it should be logical to exist a corresponding option to withdraw or revoke that consent or to make subsequent changes to that consent [31, 32]. The principle of consent withdrawal within the Human Computer Interaction (HCI) context has been studied in many ethical research projects, with Benford et al. in [33] to underline that in many cases it may be difficult to fully withdraw in practice because balancing consent, withdrawal and privacy is a very demanding managed task. Whitley in [32] argues further that, since the revocation of consent can mean a variety of different things depending on the circumstances and constitutive purposes that the data are being held for, it is helpful to differentiate between revoking *the right to hold* personal data and revoking *the right to use* personal data for particular purposes. Revoking the right to hold might be implemented by marking a particular record as no longer "being live" or may require the deletion of records and, in extreme cases, it might require deleting data from backups and physically grinding the hard disks.

Above all, however, providing auditable, privacy-friendly proof of compliance when and how the revocation has taken place throughout the supply chain is a challenge both technologically and legally. Considering that an individual revokes consent with the original service provider, this needs to pass through to all other service providers who are handling the data, whereas the original service provider and the individual need assurance that the revocation has taken place through the chain [32]. Nevertheless, the advancements towards privacy-enabled networks and infrastructures cause concerns to academics [34] who afraid that the same mechanisms put in place to protect the privacy of data (like de-identification) may actually end up to make the tracing and removing of an individual derived data (for allowing participants to withdraw their consent completely and be forgotten) to be very difficult, or even impossible. In such situations, as Kaye [34] underscores, it may be only possible to prohibit the entry of new information and samples into the system.

Apart from these practical difficulties, there are also economical and public-good arguments for disallowing absolute withdrawal. For instance, in the bio-banking field complete withdrawal could lead to the wastage of resources invested in bio-repositories [34, 35] whereas the practice of archiving qualitative research data for substantive secondary analysis can be significantly challenged under the revocation mechanism for withdrawing consent [36]. In the light of these immense consequences, the concept of consent withdrawal is largely in question by many biomedical and legal experts.

3.4.1 Consent Misuses

Notwithstanding that wide use of informed consent within the scientific and biomedical research domain, there exist plenty of concerns raised by the collection and analysis of data from potentially "unwilling" participants in other application domains which store massive amounts of personal data, such as social media platforms, smartphone applications, or open web forums [27].

Lately, cases of bad practices of obtaining consent have been observed extensively. One such example is the contagion study commenced by Facebook [37] where the company manipulated users by changing their newsfeed to investigate whether the emotional state of users could be influenced by the words of other users—a form of "emotional contagion" which takes place unbeknown to the user. The study provoked extended criticism of the Facebook research practices and resulted in its characterization as "*the best human research lab ever*"[4] [38, 39], even though the company publicly acknowledged and apologized for its fault.[5] It should be noted that in the study no user consent was ever required on the grounds that users were already given broad consent when they signed in to use the social network.

Facebook has seemingly a long history for conducting online research with its users' personal data without obtaining explicit or direct consent. Back in 2010, a company's experiment with 61 million users [40] resulted in changing the real-world voting behaviour of millions of people for the US midterm elections. As Gleibs emphasizes in [8], even though the statistical effect of the manipulation was small, their intervention might have had the potential to change the outcome of the Congressional elections in 2010, and although the study influenced behaviour with good intentions, the techniques employed could be used to influence political protest or anti-democratic behaviour in countries with little democratic traditions [8]. In the study, no consent was obtained from the participants on the grounds that the

[4] http://www.forbes.com/sites/kashmirhill/2014/06/28/facebook-manipulated-689003-users-emotions-for-science/#197c5800197c.

[5] http://www.cbsnews.com/news/researcher-apologizes-for-facebook-study-in-emotional-manipulation/.

experiment was not intrusive to people's lives, it bore minimal risk to the participants, it didn't affect their rights, and the research could not have been possible otherwise [8].

In another research study [41], a group of Danish researchers publicly released a dataset of nearly 70,000 users of the online dating site OkCupid, including usernames, age, gender, location, relationship (or sex) preference, personality traits, and answers to thousands of profiling questions used by the site.[6] Researchers excused themselves for not obtaining users' consent by stating that the data were already public. Nevertheless, as Michael Zimmer explains in [42], "*just because personal information is made available in some fashion on a social network, does not mean it is fair game to capture and release to all*".

Although the number of research projects not obtaining consent for exploiting personal data is quite large, the failures of the past (e.g. Harvard's discontinued sociology research project using Facebook sensitive data without consent and compromising participants' privacy [43]) seem to have alarmed the Institutional Review Boards when they are confronted with research projects based on data gleaned from social media [42]. Aside from research, however, and for offering suicidal users a second chance, Samaritans, a leading suicide prevention charity, launched a few years ago the Radar App, an app designed to tell Twitter users which of the people they follow might be feeling low. This was achieved by using an algorithm to identify keywords and phrases in peoples' tweets that indicated distress or a mentally vulnerable state and to notify their followers accordingly. The app provoked mass media uproar and wide criticism due to the raised data protection and privacy issues as it has been allowing the sharing of personal information with other untrusted people without the subject's knowledge or consent.[7] Finally, few weeks after its launch, the app was permanently suspended.[8]

In the medical domain, and in the NHS UK in particular [44], the case of using patient data without offering clear, specific, free and informed consent, not even unambiguous and effective opt-outs, while also misleading about the level of anonymization of their data and the likelihood of re-identification with the argument that research is part of their "care", has been considered to be a stretching of the law. Ultimately, these practices led to the closure of the UK national program Care.data which aimed at the integration of all patient data in a single platform.[9]

In recent years, the cases of misusing user's consent for profit or for political gains are more and more frequent. The latest Facebook–Cambridge Analytica political data scandal revealed in early 2018, is thus far the most notorious example of misusing and manipulating users' consent for political advertising purposes. The company

---

[6] http://www.wired.com/2016/05/okcupid-study-reveals-perils-big-data-science/.

[7] http://www.theguardian.com/society/2014/nov/07/samaritans-radar-app-suicide-watch-privacy-twitter-users.

[8] http://www.samaritans.org/how-we-can-help-you/supporting-someone-online/samaritans-radar.

[9] http://www.nationalhealthexecutive.com/Health-Care-News/nhs-england-to-close-caredata-programme-following-caldicott-review.

had harvested the personal data of more than 50 million users without their consent and used it to target individual US voters in 2016 elections. These vast amounts of data allowed the company to exploit the "psychographic profiles" of millions and to develop techniques that underpinned Trump campaign in 2016.[10,11]

The above examples are just a small fragment illustrating the chaos and uncertainty that dominate industry, academia, and public authorities in obtaining and revoking consent for using personal data. Although this gap is usually expected to be addressed by ethical guidelines and specialized policies, it is actually the territorial legislation the one that will enforce common handling against various interpretations of national policies. Therefore, the long-awaited GDPR regulation has already raised high expectations in dealing with such sensitive issues.

3.4.2 Consent Under the GDPR

Since the early years of the GDPR introduction, its imposed consent obligations for research have been extensively criticized by the academic and medical community [45–48]. Upon its final publication in 2016, extensive analysis has been conducted by law scholars, not only in Europe but worldwide, and most of the hesitations about compromising research were dropped given the "*research exemption*" anticipated by the regulation (Recital 33). In particular, the GDPR acknowledges that "*it is often not possible to fully identify the purpose of data processing for scientific research purposes at the time of data collection*" and continues that "*Data subjects should be allowed to give their consent to certain areas of scientific research when in keeping with recognized ethical standards for scientific research. Data subjects should have the opportunity to give their consent only to certain areas of research or parts of research projects to the extent allowed by the intended purpose*" (Recital 33). Moreover, the GDPR specifies derogations for research without consent in cases of medical research conducted "*in the public interest*" or for compliance with legal obligations (Recital 51). Nevertheless, when consent is to be used, consent presumed by failure to opt-out or by change pre-ticked boxes will no longer be permitted because consent should need to be provided by a "*clear, affirmative action*" (Article 4.11) [49].

Besides these derogations, overall the GDPR creates additional hurdles for consent over what was required by the DPD.[12] Particularly, the conditions for obtaining consent under the GDPR have become stricter since consent has to be, not only informed and specific, but unambiguous as well. Whereas the earliest drafts proposed by the European Commission specifically had introduced the requirement of "*explicit*" consent for processing all kinds of personal data, the final document clarifies that explicit consent is required only for processing sensitive personal data (Article 9(2)). For non-sensitive data, however, a freely given, specific, informed and

[10] http://www.venafi.com/blog/privacy-and-consent-heart-cambridge-analytica-scandal
[11] http://www.nytimes.com/2018/03/17/us/politics/cambridge-analytica-trump-campaign.html.
[12] http://iapp.org/news/a/top-10-operational-impacts-of-the-gdpr-part-3-consent/.

unambiguous consent will do, and this allows the possibility of implied consent if an individual's actions are sufficiently indicative of their agreement to processing [50]. Additionally, consent has to be easily withdrawn and not to be assumed from inaction. Inevitably, these requirements translate to the amendment of many current data protection notices. However, it has been underscored [3] that the GDPR document evidently constitutes the next-best option in order to warrant a significant level of protection as it appears relevant to the contemporary processing needs. On the other hand, since data that do not pertain to natural persons are beyond the scope of the GDPR, it is argued that it fails to protect individuals in the case of automated algorithmic decisions that do not target individuals but affect their lives [6]. As we also discuss in Sect. 6.2.4.2, this is the case of the "*tyranny of the minority*", a term introduced by Barocas and Nissenbaum [24] to describe the choice forced upon the majority of the population by a consenting minority who will disclose information about themselves but this information may implicate others who happen to share the more easily observable traits that correlate with the traits disclosed. As they explain, "*the value of a particular individual's withheld consent diminishes the more effectively one can draw inferences from the set of people that do consent, when this set approaches a representative sample. Once a dataset reaches this threshold, analysts can rely on readily observable data to draw probabilistic inferences about an individual, rather than seeking consent to obtain these details*".

Notwithstanding this deficiency, the GDPR anticipates for a right to withdraw (revoke) consent, a fact that has been warmly applauded since this explicit reference to the right to withdraw consent was missing from the DPD [31, 51]. Under this right, the data subject has the right to withdraw consent at any time, but the revocation is foreseen only for the future processing of personal data. Therefore, the data controller should not use the data for future assessments and processing, i.e. the revocation is not retroactive as it does not apply for processing that had taken place before withdrawal: "*The withdrawal of consent shall not affect the lawfulness of processing based on consent before its withdrawal*" (Article 7(3)). Hence, this notion of non-retroactive revocation enshrined in the GDPR is not affected by the progress of informational privacy infrastructures and neither devaluates already conducted research.

This also complies with the Opinion on the definition of consent [52] published in 2011 by the Article 29 Working Party (WP29),[13] which specifies that withdrawal is exercised for the future, not for the data processing that took place in the past in the period during which the data was collected legitimately. Decisions or processes previously taken on the basis of this information can therefore not be simply annulled. However, *if there is no other legal basis justifying the further storage of the data, they should be deleted by the data controller*. The GDPR Article 7, however, leaves open for interpretation whether this provision about consent affects, apart from the processing (which does not), the storage of the data themselves on which the with-

[13] The Article 29 Working Party (WP29), set up under Article 29 of Directive 95/46/EC (DPD), is an independent European advisory body on data protection and privacy bringing together the European Union's national data protection authorities. As from 2018 the Article 29 Working Party has been transitioned into a new legal framework under the GDPR, the European Data Protection Board (EDPB).

drawal applies, and therefore it does not clarify if it requires the erasure of the data upon their revocation of consent under which they were first collected. Nevertheless, following the introduction of the RtbF in Sect. 3.5.3 this issue unravels.

Supplementary to the right to revoke consent, two more powerful rights have been foreseen under the GDPR, the right to object (Article 21) and the right to restriction of processing (Article 18). Although the right to object was also specified in DPD where compelling legitimate grounds must be demonstrated by the data subject in order to object to the processing of personal data, under the GDPR the definition of the right to object is significantly expanded since the burden is put on the data controller to demonstrate compelling legitimate grounds when a data subject is objecting to the processing based on public interest (Article 6(1)(e)) or the legitimate interests of the controller (Article 6(1)(f)). By exercising the right to restriction of processing data subjects have the right to restrict the processing of personal data when the conditions specified in Article 18(1) apply, and consequently the data may only be stored by the controller, but they cannot be further processed.

3.4.3 Current Efforts for Revoking Consent

For the functional implementation of feasible consent mechanisms, many frameworks, both legal and technical, have been proposed over the past few years. An indicative portion of them is presented here.

Within the medical field, an option, from a legal perspective, for implementing informed consent efficiently is not to implement any constraints at consent at all! As oxymoron as it sounds, this concept is adopted by the Portable Legal Consent (PLC), a US legal framework for consent in research developed by the Consent to Research project.[14] The project aimed at developing a process through which individuals can make an informed choice about participating in research through the clear communication of risks, benefits, and consequences [26]. It allows participants who are willing to relinquish control of their personal information to attach a one-time research consent to their health and genetic data, which they upload themselves onto the web site [53]. Participants may withdraw their data from the database at any time, but they are clearly advised that once data are uploaded, it may not be possible to remove them from all sources (for example, from researchers who have already downloaded, shared, or used the data). This Portable Legal Consent requires participants to go through rigorous consent processes and demands honesty and trust from both researchers as well as participants [8].

Almost a decade ago, researchers, in an attempt to provide a technical solution for granting and revoking consent under the DPD requirements, proposed an approach that provides for a verifiable and revocable expression of consent and allows services to gain a proof of consent even for aggregated personal data [54]. The solution builds on a digitally signed hash tree and reuses PKI mechanisms, especially certificates and

[14] http://del-fi.org/consent.

certificate revocation, in order to cater for changes in the expression of consent and to allow the vanish of a once established consent, all accomplished without the need of a direct relationship or the iterative involvement of the data subject. However, as explained in [54], the solution does not avoid the non-consented processing of data.

Giving and revoking consent effectively has been the scope of many research projects, such as the EnCoRe (Ensuring Consent and Revocation) project, a large, cross-disciplinary research project in the UK. The project investigated how to improve the rigour with which individuals can grant and, more importantly, revoke their consent to the use, storage and sharing of their personal data by others [55]. One of the main research goals was to ensure revocation compliance throughout the supply chain, i.e. if an individual revokes consent with the original service provider, this needs to pass through the supply chain to all the other service providers who are handling the data, whereas the original service provider (and the individual) need to be assured that the revocation has taken place effectively through the chain. It was within the EnCoRe project that the notion of dynamic consent was first coined by Professor Kaye and her team [56] as a way to provide dynamic and granular options for revocation in system design [55]. As authors explain, in dynamic consent the reliable storage and enforcement of consent choices is achieved by cryptographically "wrapping" the individual's dynamic consent preferences with samples/information provided. This is possible because of machine-readable disclosure policies or "sticky policies" that are attached to data [57, 58]. Sticky policies attach metadata that define conditions and constraints describing how the data should be treated, and they are strictly associated with users' data driving access control decisions and privacy enforcement [59]. This package of "wrapped information", which contains specific consent provisions, travels with the participant's data as these are shared or accessed for different purposes [22]. Under EnCoRe pilot this "wrapped information" embraced new homomorphic encryption techniques[15] [62], which allowed information to be processed in its encrypted state while permitting the results of the processing to remain encrypted [63, 64].

Urquhart et al. in [65], acknowledging the unequivocal place of consent in the IoT era of embedded physical devices, proposed a route forward for changing how consent is obtained by using the concept of "trajectories" which already used within the HCI studies for understanding and designing complex user experiences. In their work, they are taking different elements of the trajectories framework; time, actors, space, interface, and map them onto designing consent processes that enable mechanisms for informing, obtaining and withdrawing consent.

The use of privacy agents, a dedicated software which would act as a "surrogate" of the subject and automatically manage on her behalf her personal data, has also been proposed for dealing with the management of data subject's explicit consent [66]. The recommended architecture for these "privacy agents" is based on formal

[15] Homomorphic encryption allows computations to be carried out on the ciphertext without decrypting it first and thus, the encrypted result, when decrypted, matches the result of operations performed on the plaintext. Homomorphic encryption has been employed, among others, in pilot studies for protecting the privacy of genomic information [60, 61].

(mathematical) semantics, a fact that enables the definition of the expected behaviour of privacy agents without any ambiguity and thus can make privacy rights protection more effective. The extensive use of state-of-the-art privacy agents, which enable people to configure their privacy preferences and exchange these preferences with data controllers through personal information policy exchange protocols, is analysed in [67].

As a proof of concept, researchers from the UK designed and developed recently an Apple mobile health app [68] for demonstrating the requirement for supporting informed consent and withdrawal in research projects. They implemented a custom-built module for consent—similar to the ResearchKit provided by Apple—whose functionality supports gaining informed consent, displaying template forms upon first launch, and allowing the collection of digital signatures. While the researchers provided the choice for a complete data withdrawal from the study, they designed this functionality as a multi-step process in order to avoid the situation where users withdraw data by mistake.

The approach of OPERANDO,[16] an EU funded project aiming at implementing and validating an innovative privacy enforcement framework, is to create a vault where users store their sensitive data and selectively share them with Online Service Providers (OSP). To this end, users access a dashboard which allows them to manage which OSP accesses what data and when, and easily revoke or grant access to the data. To facilitate OSPs, OPERANDO allows them to query the stored data with the use of the OData standard[17] and enforce the user's privacy policies in the results of each query.

In the same context, various web standards have emerged for the specification and implementation of consent procedures in online environments. One such standard is the User-Managed Access (UMA) [69] which has been approved by the Kantara Initiative[18]. UMA proposes an OAuth-based architecture that enables conforming applications to offer stronger consent management capabilities and an asynchronous, centralized protocol for consent. While UMA has been under development for several years, its specifications have now been stabilized and support multiple implementations and a widening variety of use cases. The authorization policies anticipated to be used in conjunction with the UMA is the eXtensible Access Control Markup Language (XACML),[19] which defines a declarative fine-grained, attribute-based access control policy language, an architecture, and a processing model describing how to evaluate access requests according to the rules defined in policies.

Kantara Initiative also supports the standardization effort of Consent Receipt,[20] a form of signed receipts in a JSON Web Token Format which can be used to improve existing consent mechanisms against the requirements specified by regulations, and

[16] http://www.operando.eu/.

[17] http://www.odata.org/.

[18] http://kantarainitiative.org/confluence/display/uma/Home.

[19] http://www.oasis-open.org/committees/xacml/.

[20] http://kantarainitiative.org/confluence/display/LC/2017/01/08/Consent+Receipt+Specification+v1.0+public+comment+and+IPR+review+period?src=contextnav.

in particular the GDPR. Consent Receipts can facilitate people's use of consent when communicating with the data controllers as well as when withdrawing consent. In conjunction with UMA mechanisms, consent receipt can be used as a tool for demonstrating effective personal control over data. According to its specifications [70], in some respects a consent receipt could be described as a reverse cookie, in that both the individual and the organization have a record of the consent, and the individual can use the receipt to track and profile the organization and/or service along with consent and information sharing preferences. Thus, people can track sharing with third parties, like third parties can track people. Inspired by the Kantara's standard, consent receipt prototypes were also deployed under the Personal Data Receipt project of Digital Catapult Centre in the UK[21] [71].

Beyond the above, there are some other efforts across the EU countries aiming at the development of nationally and internationally interoperable models for personal data management. Examples are the MyData Initiative in Finland,[22] which specifies reference architecture in order to provide a rigid framework for consent and data authorization management via a standard and interoperable mechanism,[23] and the Consent Group within the Personal Data and Trust Network in the UK,[24] which aims at using next-generation standards in consent and facilitating the use of a consent-based trust in digital framework use cases.

Evidently, since the publication of the final text of the GDPR and the new requirements brought upon the Privacy Shield agreement[25]—the framework for regulating transatlantic exchanges of personal data between the EU and US—the technical discussions on the feasibility of granting and revoking a simple, informed, unambiguous and declarative consent, along with its receipt as a proof of discourse, have been intensified. The Real Consent Workshops, which was the combined effort from big standard initiatives like Kantara and Digital Catapult, intended to delve into the gap between the type of consent people find meaningful and what we have online today.[26] Real Consent efforts were focused on both technology and policy, and they were comprised of open standards and best practices. Most importantly, Real Consent aimed at developing a collection of assets to address the current challenges around consent. In parallel, the Open Consent Framework was an approach to operationalize standard notices with a trust framework, i.e. a notary function where trusted third-party organizations can register/generate their notices and in which additional layers of technology can be added to provide more advanced and more trusted user functionality.[27]

[21] http://www.digicatapult.org.uk/news-and-views/publication/pdr-report.

[22] http://mydatafi.wordpress.com/.

[23] http://hiit.github.io/mydata-stack/.

[24] http://pdtn.org/roundup-panel-meeting-on-data-ethics/.

[25] http://www.privacyshield.gov.

[26] http://real-consent.smartspecies.com, http://kantarainitiative.org/real-consent-workshops-the-consent-tech-bubble-grows/.

[27] http://www.smartspecies.com/open-consent-framework/.

3.5 The Right to be Forgotten

Although the non-retroactive definition of consent revocation does not allow the forgetting of past processes and inferences carried out based on personal data once they were collected, the GDPR introduces the concept of the Right to be Forgotten (RtbF) or the "Right to Oblivion" for allowing the retro-active erasure of the actual personal data themselves. According to the RtbF's retro-activity, individuals have the right to request the retroactive erasure of all of their personal data. Among other conditions, the withdrawal of a previously given consent is sufficient to trigger the RtbF to allow the erasure of the personal data *from every data controller* who is processing them and not only from the one who processed the data in the first place (Article 17(2)). The fact that the consent was provided only to the original controller does not appear to be relevant for the right to take effect.

Forgetting previously collected personal data, obtained either because the user has once submitted them or because an online service has sneakily scrapped them, has been for long a disputable and controversial matter the European Commission attempted to untangle with legislation. Given the notable infeasibility for users to maintain control of their data, their diffusion and their subsequent uses once they were collected, the right aims at counterbalancing this luck of transparency on personal data processing and secondary use.

3.5.1 Forgetting and the Need to be Forgotten

The right evolves from the need for forgetting which, according to Bannon [72], is a central feature of our lives, even though it has a relatively little serious investigation in the human and social sciences. He outlines that judicious forgetting is of fundamental value both for individuals and societies, a necessary human activity and not simply a bug in the design of the human. Although we most often live under the assumption that remembering and commemorating is usually a virtue and forgetting is necessarily a failing, many scholars argue otherwise [73, 74]. As Connerton states in [73], forgetting is not always a failure, and it is not always, and not always in the same way, something about which we should feel culpable. He describes how the Ancient Greeks provide us with a prototype of prescriptive forgetting as they were acutely aware of the dangers intrinsic to remembering past wrongs because they well knew the endless chains of vendetta revenge to which this so often led. Tirosh in [74] denotes that Romans, and the nations created subsequently from the 9th to the 20th century, employed diverse practices of forgetting as means to a greater end—the process of creating a common, shared communal memory. Memory processes have always contained both the practices of forgetting and of remembering since our memory is a combination of what we remember about our past, what we may have forgotten about it, and what we wish to forget. Even within the justice system, we see the development of practices that require certain kinds of deliberate forgetting after a period in order to allow people to have a new start in life and not be haunted by an indiscretion many years earlier [72].

The importance of forgetting—whether by individuals, groups, organizations or even nations—has been observed and commented by scientists, historians, politicians, philosophers, writers and poets through the ages. French distinguished philosopher Ricoeur in "*Memory, history, forgetting*" [75] discusses the necessity of forgetting as a condition for the possibility of remembering and affirms that the "power to forget" is necessary to all actions, describing it as the very power that allows the one possessing memory and history to "*heal wounds, to replace what has been lost, to recreate broken forms out of itself alone*". In "*The End of Memory*", Volf [76] clarifies that the injunction to remember carries within itself an allowance for forgetting: "*Remember, yes; but for how long?*" he questions and extols "how to remember rightly" so that memory might be able to rest. German philosopher Friedrich Nietzsche in his essay "*On the use and abuse of history for life*" [77] demonstrates how it is generally completely impossible to live without forgetting.

As Mayer-Schönberger describes in his widely cited work [78], forgetting performs an important function in human decision-making. It allows us to generalize and abstract from individual experiences. It enables us to accept that humans, like all life, change over time. Yet, it is commonly stated that forgetting is not neutral in a political and historical perspective in which it can be seen as both a blessing and a curse. A blessing because for moving forward one needs on occasion to forget, bury the past and forge a new beginning whereas can also be a curse for those who forget the past and thus are condemned to repeat it [72]. In the medical domain, the recent research on hyperthymesia condition, an "unusual autobiographical memory" [79], found that it disrupts the lives those living with it as they describe their never stopping memories to be "exhausting" and a "burden". The condition has been the inspiration for many fiction and movie artifacts.[28] The most notable is Borges' infamous fiction book "*Funes the Memorious*" [80] that describes a man who is suffering by a similar condition as he is being haunted by his inability to forget anything, and as a result, his life has been a misery.

Although in the paper-and-ink world, as explained in [81], the sheer cumbersomeness of archiving and later finding information often implied and promoted a form of institutional forgetfulness—a situation with parallels to human memory—in the digital world our digital records constitute an array of potential memories the very existence of which may compromise our ability to forget or move on [82–84]. Whereas in the past forgetting was the default, due to the cost and rigour embroiled in remembering, digital age changed this assumption and caused the balance of remembering and forgetting to be inverted and thus forgetting to be the exception [78]. As de Andrade argues in [85], "*the past is no longer the past, but an everlasting present*", while Burkell [83] underlines that our ability to construct and maintain our own identities is threatened by digital systems that "remember" everything about us: thus, there is value in, and a need for, forgetting and being forgotten. Therefore, in the context of informational systems, we should view forgetting as a feature and, on that

[28] http://en.wikipedia.org/wiki/Hyperthymesia.

account, we should try to use technology to augment forgetting in human-computer interaction in order to "teach" computers to forget [72].

Yet, computer scientists have not given a lot of thought on the phenomenon of forgetting as the capacity of modern computers to store everything and never forget has always been considered as a want-to-have feature. However, according to Blanchette and Johnson [81], who were among the first computer scientists to have envisioned the need for forgetting within the information systems, an unintended side effect of data retention is the disappearance of social forgetfulness that allows individuals a second chance for a fresh start in life. Therefore, they argue that privacy policies must address not only collection and access to transactional information but also its timely disposal as part of a broader and comprehensive policy approach [81]. In this regard, Dodge and Kitchin [86] had been arguing almost a decade ago that rather than seeing forgetting as a weakness or fallibility, it must be seen as an emancipatory process that will free pervasive computing from burdensome and pernicious disciplinary effects.

As digital remembering undermines the important role that forgetting performs, Mayer-Schönberger argues that it threatens us individually and as a society in our capacity to learn, to reason, and to act in time [78]. For example, the sheer amount of information that computers provide and hold—too many email messages in our Inbox and too much history—can be a source of frustration due to the "information overload" phenomenon which induces a state of paralysis, affecting people's ability to act [72]. What's more, digital memories, comprising of the vast amount of data continuously collected as we go about our everyday lives, make possible a comprehensive reconstruction of our words and deeds and, even if they are long past, they strongly suggest we are moving into a panoptic society as they create a temporal version of Bentham's panopticon [87], constraining our willingness to say what we mean [78, 81, 88]. Search engines, most notably Google Search, stand at the heart of this panoptic architecture of the Internet [88] as web enables the retention of large quantities of personal micro-information over time, which can provide for an extremely detailed reflection of our past. Rosen [89] remarks that "*the fact that the Internet never seems to forget is threatening, at an almost existential level, our ability to control our identities; to preserve the option of reinventing ourselves and starting anew; to overcome our checkered pasts... The Internet is shackling us to everything that we have ever said, or that anyone has said about us, making the possibility of digital self-reinvention seem like an ideal from a distant era*". Within this context, Solove in his book [90] explains how the free flow of information on the Internet can make us less free: "*Information that was once scattered, forgettable, and localized is becoming permanent and searchable. Ironically, the free flow of information threatens to undermine our freedom in the future. These transformations pose threats to people's control over their reputations and their ability to be who they want to be. The more freedom people have to spread information online, the more likely that people's private secrets will be revealed in ways that can hinder their opportunities in the future*". Indeed, examples of the devastating consequences of digital forgetfulness are spread across the literature. Indicative cases are the teacher who lost her job over a photo of her holding a glass of wine posted on Facebook,[29] and more recently, the case of Harvard University who withdrew acceptance of 10 freshmen over to their offensive postings in a group Facebook chat.[30]

Undoubtedly, with the rising of the WEB 3.0 era [91] the explosion of data on the web has emerged as a new problem space. Semantic web technologies integrated into, or powering, large-scale web applications and Linked Data best practices for publishing and connecting structured data on the web [92] contribute to the boosting of personalization and contextualization of information. The dominance of intelligent search services and the efficient inferences produced by artificial intelligence (AI) algorithms pave the way for the endless information dissemination and the vitiation of forgetfulness. Consequently, the need for forgetting in our digital age has begun to preoccupy more and more computer scientists who realize that forgetting is an essential part of HCI systems. For instance, in recent years there have been attempts for modelling forgetting in robotic devices in order to support a more realistic and natural digital illusion of life experience [93]. Additionally, the value of intentional forgetting at situations in which people may be highly motivated to forget has been studied so as to provide implications for designing complex practices associated with problematic disposal of digital possessions [94].

3.5.2 About the CJEU Decision

Amidst the social and philosophical discussions on the criticality of forgetfulness in the digital era, in 2014 the Court of Justice of the European Union (CJEU) tried to tackle the need for forgetting through the infamous Google Spain decision which forced Google to take down harmful personal information from its search results [51, 95]. Although the final settlement ordered Google to remove the relevant link at first only from its corresponding Spanish domain and later from all the European Google sites, later the French privacy authority CNIL requested irrelevant and out-dated contents to be removed from all the non-European Google sites as well and therefore fined the company for non compliance. This provoked an intense debate between CNIL and Google which appealed to this request.[31] After a long period of waiting for a decision, in September 2019 the European Court of Justice (ECJ) ruled in Google's favour stating that "*the right to the protection of personal data is not an absolute right, but must be considered concerning its function in society and*

[29] http://www.cbsnews.com/news/did-the-internet-kill-privacy/.

[30] http://www.washingtonpost.com/news/morning-mix/wp/2017/06/05/harvard-withdraws-10-acceptances-for-offensive-memes-in-private-chat.

[31] http://www.bloomberg.com/news/articles/2017-01-25/google-argues-right-is-wrong-in-clash-with-french-privacy-czar.

be balanced against other fundamental rights, in accordance with the principle of proportionality".[32,33]

Contrary to the general impression, however, it has been pointed out by many scholars [51, 74, 95] and the Commission itself that the Google Spain judgement does not create an RtbF such as the one enshrined in the GDPR, as the CJEU could not have enforced a right that did not exist in the current legislation, but simply applies the RtbF which was already present (although not explicitly mentioned) in the existing legal framework, extending the lawfully published information right and the right to object. Still, the decision planted the seeds to affirm something that goes in the direction of the RtbF enshrined in the GDPR [51].

Although plaintiff's original intention was to remove the disputed information from the online archive where it was originally posted, CJEU ruling aimed at the technological intermediary and not the original publisher of the information and thus, the information was legally retained in the online archive whereas the links to the information removed from the Google Search. This was considered by some, such as Gorzeman and Korenhof [88], as an elegant solution: history is still retained and accessible, but the access is less easy. Forgetting invoked by the CJEU's decision may in time challenge historians with the retrieval of information in order to get an accurate view of past societies, but this difficulty is not automatically an impossibility, something that would be the case if the information were thoroughly deleted on the storage level or not encoded at all [88]. The court clarified that Google has to carefully balance the request for removing search results with of all the rights involved, including the public's right to have access to information. This would limit the application to cases only where the information to be deleted is both damning and irrelevant. As Mayer-Schönberger emphasizes in [96], search engines don't have to redesign themselves to comply since Google is already handling millions of deletion requests for copyright violations every month. Indeed, the CJEU decision has had such an impact that, at the time of writing, more than 919 446 requests have already been filed to de-list 3 615 071 urls from the Google search[34] and 46.3% of them have been satisfied.

While the CJEU decision had been accused of censorship, as O'Hara's comments in [97], the CJEU decision, although makes life complicated for the big corps, isn't targeted at particular types of information or data subjects, and therefore cannot be considered censorship. Baum [98] also explains that "*the court's ruling simply takes us back to the time when, if you wanted to find out something about someone, you had to dig for it; you had to know where to look for it. Censorship would be if the offending records themselves were expunged, and that is not what the court ruled*". Indeed, as the disputed information is not to be deleted from the web, and hence censored, the decision was ultimately not about the fundamental balance between privacy rights and expression rights when dealing with personal information over the web [74]. As the impact analysis of this decision is still ongoing and its distinction with the RtbF

[32] http://www.nytimes.com/2019/09/24/technology/europe-google-right-to-be-forgotten.html

[33] http://www.zdnet.com/article/google-wins-right-to-be-forgotten-case-in-europe/

[34] http://www.google.com/transparencyreport/removals/europeprivacy/.

under the GDPR is rather vague, many scholars argue that the relationship between the regulation's RtbF and the CJEU's reasoning will clearly require careful elaboration hereafter [3]. Nevertheless, Mayer-Schönberger underscores that the CJEU decision has not definitely solved the challenge of comprehensive digital remembering [96].

3.5.3 The Right to be Forgotten Under the GDPR

Back in 2012, the EU, in an attempt to respond to the challenges posed by digital remembering and having as an ultimate goal to give control of personal data back to individuals, put forward the RtbF in its then proposed regulation. The right evolves from the national law in many European countries like France, where the Right to Oblivion is anticipated. According to some legal experts, the RtbF enshrined in the GDPR has more a symbolic importance than a substantive effect as it does not actually represent a revolutionary change to the existing data protection regime. Instead, its roots lie within the DPD and in particular within the right to erasure and the right to object, even though the GDPR is more analytical in defining the right and the conditions under which it shall be invoked [85, 99, 100]. For instance, the condition of withdrawing consent in order for the RtbF to be triggered has not been encompassed in any national or European data protection law so far [99].

Admittedly, this right as introduced in the Article 17 of the GDPR is a breakthrough on the EU legislation domain because it does not only encompasses the right to erase (or "to forget") but it also embraces the right "to be forgotten". While the first specifies the need for a controller to delete data, the latter implies the need for data to be deleted *from all possible sources* in which they reside. According to extended legal analysis [3, 51], the right is a novelty and has a broader scope than any of the existing rights whereas its unique feature, which makes it different from the rights granted by the existing legislation, is its retro-activity. Article 17(1) provides several situations where a person has the right to ask personal data to be erased by the data controller. Of particular interest is sub-paragraph b, which allows the person to withdraw his or her consent. In other words, based on the GDPR, withdrawal of a previously given consent is sufficient to have personal data erased by the controller. Under the regulation, an individual can request the erasure of his personal data *from every data controller* who is processing the data and not only from the one who processed the data in the first place (Article 17(2)). The fact that the consent was provided only to the original controller does not appear to be relevant since the obligation for erasure arises when the person withdraws consent, without any specification on the controller who received it.

From the above, it is evident that the enforcement of this right would pose major technical issues due to the practicalities involved in knowing all the controllers who are processing the personal data in question. Even in the case where controllers do have knowledge of the third parties processing some data that they collected, it places upon them the additional obligation to inform those third parties about the erasure request, given that Article 17(2) states that "... *the controller shall take reasonable steps, including technical measures, to inform controllers which are processing the personal data that the data subject has requested the erasure by such controllers of any links to, or copy or replication of, those personal data*". Whereas the GDPR provides a convenient exemption from the obligation to inform all recipients of any rectification or erasure when this "*proves impossible or involves a disproportionate effort*" (Article 19), this exemption has also raised some concerns regarding the effectiveness of the RtbF as its scope of applicability is not always obvious [101].

In the final analysis, controllers are required to implement technical solutions to allow the tracking of personal information and to prove its efficient removal in the case of a request for erasure under the RtbF. And although the first may not be considered a difficult task, since many controllers keep links of their copied information, the burden to prove that the erasure has been implemented successfully from all available sources is still technologically questionable. The fact that the regulation does not provide a clear and unambiguous definition of the RtbF regarding its non-trivial practicalities of enforcing such a deletion when secondary uses apply, i.e. personal data have been disseminated to third parties, or they have been anonymized or pseudonymized, led many to argue that its future enforcement is reasonably doubted [102, 103].

Inevitably, the right provoked plenty heated debates, and fierce discussions within law, philosophy, social, humanitarian and computing disciplines and has been long explored in surveys, proposals and academic writings. Xanthoulis in [103], acknowledging that oblivion has been proven under certain circumstances to be a necessity for safeguarding human well-being, asserts that the RtbF should be conceptualized as a human right and more specifically as an expression of the broader right to privacy. Instead, de Andrade in [85] presents the RtbF as a branch of the right to identity, which is the right to be different, not from others but from oneself, i.e. from the one(s) we were before. Therefore, the RtbF—as part of the right to personal identity—is intimately connected to the ability to reinvent oneself, to have a second chance, to start over and present a renewed identity to the world. Given that the objective of the RtbF is, in many cases, not to conceal private information from public view, but to erase public information and to prevent its further disclosure, he argues that the RtbF should be understood as a (procedural) data protection right that mainly pursues and protects a (substantive) identity interest, operating to enforce an individual's right to personal identity [85]. Following this line, Burkell [83] explores the consequences of the digital record of our lives for our identity and how this digital record affects our ability to construct our own personal narratives which are central to a sense of "self". He concludes that the RtbF may be, above all else, a psychological necessity that is core to identity—and therefore a value that we must ensure is protected.

Yet, the RtbF has been met with intense resistance from both businesses and free speech advocates due to its collision with other rights and protected interests [3, 28, 85, 104]. They questioned the regulation's incentives, and they emphasized the difficulty on achieving a delicate balance between the involved rights, namely the right to privacy and the right to freedom of expression which, along with the right to privacy, is also contained in the European Convention on Human Rights (Article 10) [105]. Google's chief privacy counsellor remonstrated that the RtbF represents the biggest threat to the free speech and expression on the Internet[106] because it is not limited just to personal data that people provided themselves through an unambiguous consent agreement, but instead, it applies to all possible cases of personal data may be found online[35] [104]. Further, the RtbF has also been labelled as censorship and disastrous for the freedom of expression[36] whereas some argued that "*a Right to be Forgotten is about extreme withdrawal, and in its worse guise can be an antisocial, nihilist act*" and "*is a figment of our imaginations as it neglects the role society plays in individual's life*".[37] At an informational level, scientists have pointed out that enforcing the RtbF would lead to preventive actions like anonymization of databases per default, something that would cause an unacceptably high amount of information loss [107].

Along the same lines, Rosen [106] condemned any efforts at regulating the Internet, and search engines in particular, as he asserted that any kind of regulation of the Internet violates the inherent code of its freedom and thus, the RtbF will bring chilling effects on the Internet era which "*will not be as free and open upon the application of the right*". On the other side, scholars and privacy experts argue that free speech is already being selected and restricted by search engines themselves [108, 109]. Google's global privacy counsel argues that "*history should be remembered, not forgotten, even if it's painful. Culture is memory*"[38] whereas other eminent theorists state otherwise [75, 76, 103], i.e., that forgetting is a necessity for the evolvement of history remembering considering that cultures seem to have been built over the course of time through a process of selective remembering and forgetting, not through total remembering. In a more compromised approach, Mitrou and Karyda point out [110] that, while the RtbF cannot be synonymous with a right of a total erasure of history, the interests of social and historical inquiry do not legitimize keeping every piece of personal information regardless the rights and interests of the persons affected.

With regard to the conflict between privacy and freedom of speech, Solove in his notable work [90] argues that "*we must protect privacy to ensure that the freedom of the Internet doesn't make us less free... we must balance the protection of privacy against freedom of speech*" and "*Both are essential to our freedom. Freedom of speech is a fundamental value, and protecting it is of paramount importance. Yet, privacy often furthers the same ends as free speech. If privacy is sacrificed at the altar of free speech, then some of the very goals justifying free speech might be undermined*" [90]. And Lindsay explains in [111] that "*privacy is not necessarily the opposite of freedom of expression—if people feel assured they have some control over their information, they are more likely to share it. On the other hand, if people know that what they say and do online will be accessible to all, and for all time, they may be more likely to self-censor*".

---

[35] http://peterfleischer.blogspot.gr/2012/01/right-to-be-forgotten-or-how-to-edit.html.

[36] http://www.wired.co.uk/article/right-to-be-forgotten-blog.

[37] http://www.theguardian.com/commentisfree/libertycentral/2011/mar/18/forgotten-online-european-union-law-internet.

[38] http://peterfleischer.blogspot.gr/2011/03/foggy-thinking-about-right-to-oblivion.html.

Other academics [88, 112] proposed a more conciliatory position, arguing that since the RtbF draws more heavily on the mechanisms of human forgetting which provides for a big greyscale (in contrast to erasure which is black-and-white), an individual can have a need for different grades of forgetting rather than plain erasure, and therefore the RtbF, instead of the plain erasure of information, could rely on the level of encoding or retrieval of the information. Nevertheless, although this position it may be considered by some as censorship at the level of information retrieval [88] and hence the least heavy yet most effective means to get the minimum amount of censorship overall, it still resembles the decision concluded by the CJEU to remove the links and not the information itself, and therefore it cannot be considered as an actual forgetting.

All in all, a large part of academia considers the RtbF to be a highly qualified right as it attempts to restore some balance in favour of individuals by providing tools for controlling their personal data, while at the same time certain conditions are foreseen to be satisfied in order the right to be applicable, such as when data are no longer needed or where data are collected or processed with a person's consent, and that consent is later withdrawn. The RtbF is also subject to important exemptions and safeguards, such as the cases where it may be in conflict with the freedom of expression and information, or for journalistic purposes and the purposes of academic, artistic or literary expression (Article 17(3) and 85). Claims that the RtbF, as introduced in the GDPR, will stifle the press are therefore untrue since there is an expressed exemption for journalists, as well as an exemption for individuals engaged in purely personal or household activities [111]. In this respect, Mantelero in [99] highlights that the oppositions the RtbF received concerning the suppression of freedom of speech represent a sort of paradox. On the one hand, big IT companies are trying to promote the idea that sharing information is a social norm and that privacy and forgetting are outdated concepts, but on the other hand, the same companies are progressively collecting an enormous amount of data in order to profile individuals and, above all, to extract predictive information with high economic, social, political and strategic value. He stresses that "*in a world where it is assumed that no value is attributed to privacy and oblivion, the only ones to gain from this abandonment of old rights are the owners of these platforms or services which have an exclusive and comprehensive view of the entire mass of data*".

Lastly, another broad area of criticism against the establishment of the RtbF comes from the fact that it may impose considerable obstacles in data transfer between the EU and third countries. As a first step for resolving the impending implementation and interoperability issues resulting from the enforcement of the RtbF, Ambrose in [113] analysed the options the non-EU countries and data controllers (like the USA) have in order to react to the establishment of such a right, while Bennett in [114] discussed how a reconciliation between the USA and the EU on the RtbF could be achieved. On this matter, Voss and Castets-Renard proposed a coherent worldwide taxonomy of the RtbF [100] in order to identify its various forms within different countries and to measure the extent to which there is a convergence of legal rules internationally. Ultimately, the Privacy Shield Framework[39] between the EU and the USA will have to deal with this issue drastically.

References

1. European Union, Regulation (EU) 2016/679 of the European Parliament and of the Council of 27 April 2016 on the protection of natural persons with regard to the processing of personal data and on the free movement of such data, and repealing Directive 95/46/EC (General Data Protection Regulation). Off. J. Eur. Union **L119**, 1–88 (2016)
2. Data Protection Directive, Directive 95/46/EC of the European Parliament and of the Council of 24 October 1995 on the protection of individuals with regard to the processing of personal data and on the free movement of such data. Off. J. Eur. Union **L281**, 31–50 (1995)
3. P. de Hert, V. Papakonstantinou, The new general data protection regulation: still a sound system for the protection of individuals? Comput. Law Secur. Rev. **32**(2), 179–194 (2016)
4. A. Cavoukian, Privacy by design—The 7 foundational principles (2011)
5. M. Langheinrich, Privacy by design-principles of privacy-aware ubiquitous systems, in *International Conference on Ubiquitous Computing* (Springer, 2001), pp. 273–291
6. M. Oostveen, K. Irion, The golden age of personal data: How to regulate an enabling fundamental right?, in *Personal Data in Competition, Consumer Protection and Intellectual Property Law*. (Springer, 2018), pp. 7–26
7. L. Edwards, Privacy, security and data protection in smart cities: a critical EU law perspective. Eur Data Prot L Rev **2**, 28 (2016)
8. I.H. Gleibs, Turning virtual public spaces into laboratories: thoughts on conducting online field studies using social network sites. Anal. Soc. Issues Public Policy **14**(1), 352–370 (2014)
9. P.D. Reynolds, *Ethical Dilemmas and Social Science Research* (Jossey-Bass Inc Pub, San Francisco, USA, 1979)
10. B. Hofmann, Broadening consent—And diluting ethics? J. Med. Ethics **35**(2), 125–129 (2009)
11. J.P. Ioannidis, Informed consent, big data, and the oxymoron of research that is not research. Am. J. Bioethics **13**(4), 40–42 (2013)
12. M.A. Rothstein, A.B. Shoben, An unbiased response to the open peer commentaries on "does consent bias research?". Am. J. Bioethics **13**(4), W1–W4 (2013)
13. F. Stevenson, N. Lloyd, L. Harrington, P. Wallace, Use of electronic patient records for research: views of patients and staff in general practice. Family Practice **30**(2), 227–232 (2012)
14. M. Sheehan, Can broad consent be informed consent? Public Health Ethics **4**(3), 226–235 (2011)
15. K.S. Steinsbekk, B.K. Myskja, B. Solberg, Broad consent versus dynamic consent in biobank research: is passive participation an ethical problem? Eur. J. Hum. Gen. **21**(9), 897–902 (2013)
16. J. Katz, Informed consent-must it remain a fairy tale. J. Contemporary Health Law Policy **10**, 69–91 (1994)
17. C.M. Simon, J. L'heureux, J.C. Murray, P. Winokur, G. Weiner, E. Newbury, L. Shinkunas, B. Zimmerman, Active choice but not too active: public perspectives on biobank consent models. Gen. Med. **13**(9), 821–831 (2011)
18. B. Brown, A. Weilenmann, D. McMillan, A. Lampinen, Five provocations for ethical HCI research, in *Proceedings of the 2016 CHI Conference on Human Factors in Computing Systems* (ACM, 2016), pp. 852–863
19. E.C. Hayden, A broken contract. Nature **486**(7403), 312–314 (2012)

[39] See footnote [40].

20. M. Mostert, A.L. Bredenoord, M.C. Biesaart, J.J. van Delden, Big Data in medical research and EU data protection law: challenges to the consent or anonymise approach. Eur. J. Hum. Gen. **2**, 956–960 (2015)
21. P. Bernal, Collaborative consent: harnessing the strengths of the internet for consent in the online environment. International Rev. Law Comput. Technol. **24**(3), 287–297 (2010)
22. J. Kaye, E.A. Whitley, D. Lund, M. Morrison, H. Teare, K. Melham, Dynamic consent: a patient interface for twenty-first century research networks. Eur. J. Hum. Gen. **23**(2), 141–146 (2015)
23. T. Ploug, S. Holm, Meta consent: a flexible and autonomous way of obtaining informed consent for secondary research. BMJ: Br. Med. J. **350** (2015)
24. S. Barocas, H. Nissenbaum, Big data's end run around procedural privacy protections. Commun. ACM **57**(11), 31–33 (2014)
25. F.H. Cate, V. Mayer-Shönberger, Notice and consent in a world of Big Data. Int. Data Privacy Law **3**(2), 67–73 (2013)
26. J. Hemerly, Public policy considerations for data-driven innovation. Computer **46**(6), 25–31 (2013)
27. B.D. Mittelstadt, L. Floridi, The ethics of big data: current and foreseeable issues in biomedical contexts. Sci. Eng. Ethics **22**(2), 303–341 (2016)
28. O. Tene, J. Polonetsky, Big data for all: Privacy and user control in the age of analytics. Nw. J. Tech. Intell. Prop. **11**, xxvii
29. E. Luger, T. Rodden, An informed view on consent for UbiComp, in *Proceedings of the 2013 ACM International Joint Conference on Pervasive and Ubiquitous Computing* (ACM, 2013), pp. 529–538
30. A. Morrison, D. McMillan, M. Chalmers, Improving consent in large scale mobile hci through personalised representations of data, in *Proceedings of the 8th Nordic Conference on Human-Computer Interaction: Fun, Fast, Foundational* (ACM, 2014), pp. 471–480
31. L. Curren, J. Kaye, Revoking consent: a 'blind spot' in data protection law? Comput. law Secur. Rev. **26**(3), 273–283 (2010)
32. E.A. Whitley, Informational privacy, consent and the "control" of personal data. Inf. Secur. Tech. Rep. **14**(3), 154–159 (2009)
33. S. Benford, C. Greenhalgh, B. Anderson, R. Jacobs, M. Golembewski, M. Jirotka, B.C. Stahl, J. Timmermans, G. Giannachi, M. Adams et al., The ethical implications of HCI's turn to the cultural. ACM Trans. Comput.-Hum. Interact. (TOCHI) **22**(5), 24 (2015)
34. J. Kaye, The tension between data sharing and the protection of privacy in genomics research. Annu. Rev. Genom. Human Gen. **13**, 415–431 (2012)
35. S. Holm, Withdrawing from research: a rethink in the context of research biobanks. Health Care Anal. **19**(3), 269 (2011)
36. O. Parry, N.S. Mauthner, Whose data are they anyway? Practical, legal and ethical issues in archiving qualitative research data. Sociology **38**(1), 139–152 (2004)
37. A.D. Kramer, J.E. Guillory, J.T. Hancock, Experimental evidence of massive-scale emotional contagion through social networks. Proc. Natl. Acad. Sci. **111**(24), 8788–8790 (2014)
38. J. Jouhki, E. Lauk, M. Penttinen, N. Sormanen, T. Uskali, Facebook's emotional contagion experiment as a challenge to research ethics. Media Commun. **4**(4), 75–85 (2016)
39. R. Schroeder, Big Data and the brave new world of social media research. Big Data Soc. **1**(2), 2053951714563194 (2014)
40. R.M. Bond, C.J. Fariss, J.J. Jones, A.D. Kramer, C. Marlow, J.E. Settle, J.H. Fowler, A 61-million-person experiment in social influence and political mobilization. Nature **489**(7415), 295–298 (2012)
41. E.O. Kirkegaard, J.D. Bjerrekær, The OKCupid dataset: a very large public dataset of dating site users. Open Differ. Psychol. **46** (2016)
42. M. Zimmer, "But the data is already public": on the ethics of research in Facebook. Ethics Inf. Technol. **12**(4), 313–325 (2010)
43. K. Lewis, J. Kaufman, M. Gonzalez, A. Wimmer, N. Christakis, Tastes, ties, and time: a new social network dataset using Facebook.com. Social Netw. **30**(4), 330–342 (2008)

44. I. Brown, L. Brown, D. Korff, Using NHS patient data for research without consent. Law Innov. Technol. **2**(2), 219–258 (2010)
45. F. Pelliccia, G. Rosano, Medical research could soon be jeopardized by new European union data protection regulations. Euro. Heart J. **35**(23), 1503–1504 (2014)
46. M. Ploem, M. Essink-Bot, K. Stronks, Proposed EU data protection regulation is a threat to medical research. BMJ **346** (2013)
47. P. Quinn, A.K. Habbig, E. Mantovani, P. De Hert, The data protection and medical device frameworks-obstacles to the deployment of mHealth across Europe? Eur. J. Health Law **20**(2), 185–204 (2013)
48. G. Rosano, F. Pelliccia, C. Gaudio, A.J. Coats, The challenge of performing effective medical research in the era of healthcare data protection. Int. J. Cardiology **177**(2), 510–511 (2014)
49. J.M.M. Rumbold, B. Pierscionek, The effect of the General Data Protection Regulation on medical research. J. Med. Internet Res. **19**(2) (2017)
50. P. Lee, K. Pickering, The general data protection regulation: a myth-buster. J. Data Protect. Privacy **1**(1), 28–32 (2016)
51. C. Bartolini, L. Siry, The right to be forgotten in the light of the consent of the data subject. Comput. Law Secur. Rev. **32**(2), 218–237 (2016)
52. Article 29 Data Protection Working Party, *Opinion 15/2011 on the Definition of Consent. WP 187*. https://ec.europa.eu/justice/article-29/documentation/opinion-recommendation/files/2011/wp187_en.pdf (2011)
53. E. Vayena, A. Mastroianni, J. Kahn, Caught in the web: informed consent for online health research. Sci. Transl. Med. **5**(173), 173fs6 (2013)
54. H.C. Pöhls, Verifiable and revocable expression of consent to processing of aggregated personal data, in *International Conference on Information and Communications Security* (Springer, 2008), pp. 279–293
55. E.A. Whitley, N. Kanellopoulou, Privacy and informed consent in online interactions: evidence from expert focus groups, in *International Conference on Information Systems (ICIS)* (Association for Information Systems, 2012)
56. J. Kaye, L. Curren, N. Anderson, K. Edwards, S.M. Fullerton, N. Kanellopoulou, D. Lund, D.G. MacArthur, D. Mascalzoni, J. Shepherd et al., From patients to partners: participant-centric initiatives in biomedical research. Nat. Rev. Gen. **13**(5), 371–376 (2012)
57. G. Karjoth, M. Schunter, M. Waidner, Platform for enterprise privacy practices: privacy-enabled management of customer data, in *International Workshop on Privacy Enhancing Technologies* (Springer, 2002), pp. 69–84
58. S. Pearson, M. Casassa-Mont, Sticky policies: an approach for managing privacy across multiple parties. Computer **44**(9), 60–68 (2011)
59. M.C. Mont, S. Pearson, P. Bramhall, Towards accountable management of identity and privacy: sticky policies and enforceable tracing services, in *Proceedings of 14th International Workshop on Database and Expert Systems Applications, 2003* (IEEE, 2003), pp. 377–382
60. E. Ayday, J.L.. Raisaro, J.P. Hubaux, Privacy-enhancing technologies for medical tests using genomic data. Technical Report (2012)
61. Y. Erlich, A. Narayanan, Routes for breaching and protecting genetic privacy. Nat. Rev. Gen. **15**(6), 409–421 (2014)
62. C. Stuntz, What is homomorphic encryption, and why should I care. Craig Stuntz Weblog (2010)
63. C. Gentry et al., Fully homomorphic encryption using ideal lattices. STOC **9**, 169–178 (2009)
64. D. Micciancio, A first glimpse of cryptography's holy grail. Commun. ACM **53**(3), 96 (2010)
65. L. Urquhart, T. Rodden, New directions in information technology law: learning from human-computer interaction. Int. Rev. Law Comput. Technol. **31**(2), 150–169 (2017)
66. D. Le Métayer, S. Monteleone, Automated consent through privacy agents: legal requirements and technical architecture. Comput. Law Secur. Rev. **25**(2), 136–144 (2009)
67. S. Spiekermann, A. Novotny, A vision for global privacy bridges: technical and legal measures for international data markets. Comput. Law Secur. Rev. **31**(2), 181–200 (2015)

68. J. Rooksby, P. Asadzadeh, A. Morrison, C. McCallum, C. Gray, M. Chalmers, Implementing ethics for a mobile app deployment, in *Proceedings of the 28th Australian Conference on Computer-Human Interaction* (ACM, 2016), pp. 406–415
69. E. Maler, Extending the power of consent with user-managed access: a standard architecture for asynchronous, centralizable, internet-scalable consent, in *Security and Privacy Workshops (SPW)*. (IEEE, 2015), pp. 175–179
70. M. Lizar, D. Turner, *Consent Receipt Specification, Version 1.1.0*. https://kantarainitiative.org/file-downloads/consent-receipt-specification-v1-1-0/ (2018)
71. T.C. Styliari , M. Nati, Researching the transparency of personal data sharing: designing a concert receipt. Digital Catapult (2016)
72. L.J. Bannon, Forgetting as a feature, not a bug: the duality of memory and implications for ubiquitous computing. CoDesign **2**(01), 3–15 (2006)
73. P. Connerton, Seven types of forgetting. Memory Stud. **1**(1), 59–71 (2008)
74. N. Tirosh, Reconsidering the "Right to be forgotten"—Memory rights and the right to memory in the new media era. Media Culture Soc. **39** (2015)
75. P. Ricoeur, *Memory, History, Forgetting* (University of Chicago Press, 2004)
76. M. Volf, *The End of Memory: Remembering Rightly in a Violent World* (Wm. B. Eerdmans Publishing, 2006)
77. F. Nietzsche, *On the Use and Abuse of History for Life* (1874)
78. V. Mayer-Shönberger, *Delete: The Virtue of Forgetting in the Digital Age* (Princeton University Press, 2011)
79. E.S. Parker, L. Cahill, J.L. McGaugh, A case of unusual autobiographical remembering. Neurocase **12**(1), 35–49 (2006)
80. J.L. Borges, Funes, the memorious, in *Avon Modern Writing No. 2* (Avon Books, 1954)
81. J.F. Blanchette, D.G. Johnson, Data retention and the panoptic society: the social benefits of forgetfulness. Inf. Soc. **18**(1), 33–45 (2002)
82. A.L. Allen, Dredging up the past: lifelogging, memory, and surveillance. Univ. Chicago Law Rev. **75**(1), 47–74 (2008)
83. J.A. Burkell, Remembering me: big data, individual identity, and the psychological necessity of forgetting. Ethics Inf. Technol. **18**(1), 17–23 (2016)
84. M. Hand, Persistent traces, potential memories: smartphones and the negotiation of visual, locative, and textual data in personal life. Convergence **22**(3), 269–286 (2016)
85. N.N.G. de Andrade, Oblivion: the right to be different from oneself: re-proposing the right to be forgotten, in *The Ethics of Memory in a Digital Age* (Springer, 2014), pp. 65–81
86. M. Dodge, R. Kitchin, "Outlines of a world coming into existence": pervasive computing and the ethics of forgetting. Environ. Plan. B: Plan. Des. **34**(3), 431–445 (2007)
87. J. Bentham, *Panopticon or the Inspection House* vol 2 (Payne, London, 1791)
88. L. Gorzeman, P. Korenhof, Escaping the panopticon over time. Philos. Technol. **30**(1), 73–92 (2017)
89. J. Rosen, *The Web Means the End of Forgetting*. http://www.nytimes.com/2010/07/25/magazine/25privacy-t2.html (2010)
90. D.J. Solove, *The Future of Reputation: Gossip, Rumor, and Privacy on the Internet* (Yale University Press, 2007)
91. J. Hendler, Web 3.0 emerging. Computer **42**(1) (2009)
92. C. Bizer, T. Heath, T. Berners-Lee, Linked data-the story so far, in *Semantic Services, Interoperability and Web Applications: Emerging Concepts*, pp. 205–227
93. C. Gurrin, H. Lee, J. Hayes, iForgot: a model of forgetting in robotic memories, in *5th ACM/IEEE International Conference on Human-Robot Interaction (HRI)* (IEEE, 2010), pp. 93–94
94. C. Sas, S. Whittaker, Design for forgetting: disposing of digital possessions after a breakup, in *Proceedings of the SIGCHI Conference on Human Factors in Computing Systems* (ACM, 2013), pp. 1823–1832
95. S. Kulk, F.Z. Borgesius, Google Spain v. González: did the court forget about freedom of expression. Eur. J. Risk Reg. **5**, 389 (2014)

96. V. Mayer-Shönberger, Omission of search results is not a "right to be forgotten" or the end of google. Guardian **13** (2014)
97. K. O'Hara, The right to be forgotten: The good, the bad, and the ugly. IEEE Internet Comput. **19**(4), 73–79 (2015)
98. R.M. Baum, *It's Not Censorship*. http://cen.acs.org/articles/92/i22/s-Censorship.html (2014)
99. A. Mantelero, The EU proposal for a general data protection regulation and the roots of the "right to be forgotten". Comput. Law Secur. Rev. **29**(3), 229–235 (2013)
100. W.G. Voss, C. Castets-Renard, Proposal for an international taxonomy on the various forms of the "right to be forgotten": a study on the convergence of norms. Colorado Technol. Law J. **14**(2), 281–344 (2016)
101. European Data Protection Supervisor, *Opinion of the EDPS on the Data Protection Reform Package*. https://edps.europa.eu/sites/edp/files/publication/12-03-07_edps_reform_package_en.pdf (2012)
102. B.J. Koops, Forgetting footprints, shunning shadows: a critical analysis of the "right to be forgotten" in big data practice. SCRIPTed **8** (2011)
103. N. Xanthoulis, The right to oblivion in the information age: a human-rights based approach. US-China Law Rev. **10**, 84 (2013)
104. J. Ausloos, The "right to be forgotten"—worth remembering? Computer Law Secur. Rev. **28**(2), 143–152 (2012)
105. European Convention on Human Rights, Convention for the protection of human rights and fundamental freedoms (European convention on human rights, as amended) (ECHR) (1950)
106. J. Rosen, The right to be forgotten. Stan. L. Rev. Online **64**, 88 (2011)
107. B. Malle, P. Kieseberg, E. Weippl, A. Holzinger, The right to be forgotten: towards machine learning on perturbed knowledge bases, in *International Conference on Availability, Reliability, and Security* (Springer, 2016), pp. 251–266
108. D.C. Nunziato, The death of the public forum in cyberspace. Berkeley Technol. Law J. **20**, 1115–1757 (2005)
109. A.H. Stuart, Google search results: buried if not forgotten. NCJL Tech. **15**, 463 (2013)
110. L. Mitrou, M. Karyda, EU's data protection reform and the right to be forgotten: a legal response to a technological challenge? in *5th International Conference of Information Law and Ethics 2012* (2012)
111. D. Lindsay, The *"Right to be Forgotten" Is Not Censorship*. http://www.monash.edu/news/opinions/the-right-to-be-forgotten-is-not-censorship (2012)
112. P. Korenhof, Forgetting bits and pieces: an exploration of the right to be forgotten in online memory process, in *Tilburg Institute for Law and Technology Working Paper Series*, vol. 4, issue 6 (2013)
113. M.L. Ambrose, Speaking of forgetting: analysis of possible non-EU responses to the right to be forgotten and speech exception. Telecommun. Policy **38**(8), 800–811 (2014)
114. S.C. Bennett, The right to be forgotten: reconciling EU and US perspectives. Berkeley J. Int'l L **30**, 161 (2012)

Chapter 4
The "Right to Be Forgotten" in the GDPR: Implementation Challenges and Potential Solutions

Abstract The GDPR, being a legal document, follows a technology-agnostic approach so as not to bind the provisions of the law with current trends and state-of-the-art technologies in computer science and information technology. Yet, the technical challenges of aligning modern systems and processes with the GDPR provisions, and mainly with the Right to be Forgotten (RtbF), are numerous and in most cases insurmountable. To this end, in this Chapter we discuss the challenges of implementing the RtbF on contemporary information systems, and we assess technical methods, architectures, and frameworks—existing either in corporate or academic environments—in terms of fulfilling the technical practicalities for effectively integrating the new forgetting requirements into current computing infrastructures. We also discuss the GDPR forgetting requirements in respect to their impact on the backup and archiving procedures stipulated by the modern security standards. In this context, we examine the implications of erasure requests on current IT backup systems, and we highlight a number of envisaged organizational, business and technical challenges pertained to the widely known backup standards, data retention policies, backup mediums, search services, and ERP (Enterprise Resource Planning) systems.

4.1 Introduction

Undeniably, the enforcement of the GDPR in 2018 has put extra burdens to data controllers operating within the EU. However, despite the long-lasting heavy discussions, negotiations and revisions on the final GDPR text and the ample time given to organization to apply the corresponding changes to their processes, products and services, few organization are yet able to prove actual GDPR compliance. In addition, the GDPR provides little if any technical guidance to the entities that are obliged to implement it. As a matter of fact, the GDPR, being a legal document, it does not specify explicitly any technical methods for data controllers to adapt their internal processes to its provisions. Legislators deliberately avoided the idea of recommending specific technical frameworks or privacy preserved methods for implementing the legal requirements introduced by the GDPR. Instead, they followed a technology-agnostic approach by specifying the functional requirements in a highly abstracted

E. Politou et al., *Privacy and Data Protection Challenges in the Distributed Era*, Learning and Analytics in Intelligent Systems 26,
https://doi.org/10.1007/978-3-030-85443-0_4

level, as far as their underlying implementation is concerned, and as such, they did not bind the provisions of the law with current trends and state-of-the-art technologies in computer science. The ultimate purpose of this approach was to allow the GDPR's adjustment to future technological innovations. Yet the GDPR's enforcement across the EU mandated businesses and organizations to align with its requirements in a transparent and efficient manner. Nevertheless, the technical challenges of conforming modern systems and processes to the GDPR provisions, and mainly to the Right to be Forgotten (RtbF), are numerous and in most cases not yet foreseen.

Beyond other challenges, the exercise of the RtbF by individuals who request the forgetting of their personal information has also become a thorny issue when applied to backups and archives. To this end, in this Chapter, we initially discuss the challenges of implementing the RtbF on contemporary information systems, and we discuss the GDPR forgetting requirements in respect to their impact on the backup and archiving procedures stipulated by the modern security standards. We specifically examine the implications of erasure requests on current backup systems and IT standards, and we analyse a number of envisaged organizational, business and technical challenges pertained to the widely known backup standards, data retention policies, backup mediums, search services, and ERP systems.

Furthermore, we evaluate existing methods, frameworks and architectures against the feasible implementation of the RtbF into current computing infrastructures. On this account, we discuss some state-of-the-art technologies and frameworks, existing either in business or academic environments, and we highlight their weaknesses and strengths in reference to user's full control over their personal data, and particularly their effective erasure from third-party controllers to whom the data have been communicated or disseminated.

4.2 Implementation Challenges

The implementation of the RtbF in the digital environment is not a straightforward task and can't be achieved without affecting the value of already collected data stores. Technically speaking, the effective implementation of Article 17(2), which require controllers to take "reasonable steps, including technical measures" to inform third parties when a data subject has requested the erasure of previously published personal data relating to them, may be proved burdensome or even impossible in any number of scenarios [1]. For instance, as the GDPR dictates, the triggering events for the erasure to take place would be the invoking of the RtbF under any of the conditions (a–f) described in the Article 17(1), which also includes the case where data subjects exercises their right to withdraw a previously given consent.

While, beyond any doubt, academia and industry are steadily working vigorously towards the design of technical solutions and the conformance of current infrastructures to the new requirements for forgetting within digital environments, existing frameworks need to be evaluated for compliance with the RtbF and, if needed, to be amended accordingly. Unfortunately though, recent exercises have demonstrated that state-of-the-art technologies used in large cloud mainframes face technical constraints which may affect the lawful implementation of the RtbF. For example, it has been underlined in an exercise regarding the compliance of Data Lake Enterprise Architecture Model with the GDPR [2] that the immutability of Hadoop is a phenomenon which does not allow files to be physically updated or deleted. Instead, a new instance of the file is created and automatically becomes an active one whereas previous instances of the files are not deleted, only flagged as not active. Inevitably, this property of the Hadoop Distributed File System (HDFS) to keep always undeleted its files, prevents Data Lake architecture from achieving compliance with the RtbF which requires assured deletion of personal data.

In the banking sector, the fact that GDPR's enforcement coincided with the integration of the 2nd Payment Services Directive (PSD2)[1] perplexes further the issues for financial institutions. In this regard, Account Servicing Payment Service Providers (ASPSP) (e.g. credit institutions, banks) have to allow Third Party Payment Service Providers (TPPs) (e.g. a Payment Initiation Service Provider (PISP) or an Account Information Service Provider (AISP)) to access the payment account of a Payment Service User (PSU). On the one hand, the requirement for Strong Customer Authentication dictated under the Article 97 of PSD2 and the Regulatory Technical Standards (RTS) of the European Banking Authority (EBA), arguably require ample user data and device fingerprinting methods for its implementation. On the other hand, however, this intense data collection and processing may risk GDPR compliance by challenging its forgetting obligations. In this respect, in 6.2.5.3 we discuss further the conflicts between the PSD2 and the data protection and privacy rights.

Biometric authentication technology also presents some conflicts with the RtbF. Whilst over the past few years there has been a surge in the use of biometric authentication which eliminates the need to remember passwords, this convenience is the biggest advantage and, at the same time, the Achilles' heel of these methods. While users do not need to remember anything and can use their fingerprints, iris, gait etc. to authenticate, they cannot replace these biometrics once they are lost. To this end, there is an increasing need for privacy-preserving schemes that will protect users' privacy. However, biometric measurements are subject to noise in the sense that each time a measurement is made some alterations are expected to occur, making thus each measurement distinct and different from its stored template. These alterations, stemming from motion blurring, divergences in luminosity, angle or other crucial factors, render the traditional cryptographic methods for private equality testing useless. To cater for this deficiency, many protocols investigating the concept of *Privacy-Preserving Biometric Authentication* have emerged recently [3–8]. They mainly exploit properties of partial and somewhat homomorphic encryption to hide the biometric measurements of the user which is to be authenticated and allow only matching operations against the template measurement to be performed. Nevertheless, these methods require the user authentication entity to have a stored copy of users' biometric measurements, a highly sensitive piece of information that cannot be forgotten when is to be used for authentications purposes. The use of cancelable biometrics, where a biohash and

[1] https://ec.europa.eu/info/law/payment-services-psd-2-directive-eu-2015-2366_en.

not the actual measurement is provided to the authentication entity [9–11], might be proved a reasonable solution to the problem of implementing forgetting under the GDPR obligations. Nevertheless, cancelable biometrics beat the principal purpose of using biometrics in the first place as users have to either carry an additional tag or remember a password.

Another tension arising from the alignment of current services with the GDPR, is the conflict between modern device interfaces, which are very intuitive and tailored to user needs, and the regulation's requirements for implementing a simple, specific, unambiguous and, in the case of sensitive data, informed consent. As currently a vast amount of individuals' information is collected to personalize user experience, the GDPR, by obliging users to consent to all and every piece of their identifying information when their personal data are collected, may hinder this practice. Apart from the case for informed consent, the requirements for consent in the DPD were very similar to those proposed by the GDPR. Yet, the high sanctions now imposed by the regulation constitutes these requirements not only mandatory to all operating data controllers but also remarkably costly if not implemented. Conforming, however, to the GDPR's consent strict requirements may be proved not only extremely cumbersome in terms of user experience but even highly critical for the quality of the service if the relevant personal information is not provided. Notwithstanding this contradiction, recent research efforts for providing informed notices in user-friendly and meaningful design choices while conforming to data protection legislations seem to be particularly promising towards overcoming this obstacle [12].

Last but not least, one of the most profound difficulties faced by data controllers is the enhancement of existing backup procedures in order to meet the GDPR forgetting requirements. Due to the reliance on the ICT, institutions are obliged to keep regular backups of their data in case of security incidents or physical disasters. A big question arising under the GDPR's forgetting legislation is how organizations should handle their backups once a user requests to remove her data. Apparently, according to the GDPR this deleting action must be performed in the backups as well, opening thus the door to potential data abuses, deliberate exploitations, or even accidental mistakes. Propagating the required erasure mechanisms to backups, empowers users and financial institutions to manipulate data integrity according to their needs, like hiding transactions from audit controls when deemed necessary. Depending on the organization policies and legal framework, user data records may have to be kept in non-volatile storage. Therefore, once a user requests the deletion of her data, non-automated—and contrary to the legal framework within the institution operates—actions have to be performed, leading to additional costs and possible legal deadlocks. Such issues may become more evident in financial institutions where records must always follow the information reliability, integrity and transparency principles. Acknowledging the huge impact of the GDPR on current backup and archiving processes, in what follows we delve into this topic in more detail.

4.3 The Impact of the GDPR on Backups and Archives

As it has been previously discussed, the introduction of the RtbF—which allows the retroactive erasure of one's personal data upon her request and from all available places to which they have been disseminated—caused prolonged controversies due to its pivotal impact on current data processing procedures and its unavoidable conflicts with other rights such as the right to free speech and the freedom of information. Of particular interest are the immense implications of the RtbF for the backup and archiving processes taking place within each organizational unit that handles personal data. Notably, already well established backup and archiving procedures specified by state-of-the-art security models are affected significantly from the GDPR erasure requirements. In this regard, we analyse hereafter the consequences of the RtbF implementation on the physical and cloud backup procedures along with its impact on the currently wide spread protocols and standards adopted in the design of most contemporary systems and frameworks.

4.3.1 GDPR Provisions for Backups and Archives

Although the GDPR Article 17(1) provides several situations where a person has the right to request personal data to be erased, the Article 17(3) of the GDPR allows for some exemptions from the "forgetting" requirement, e.g. for cases of "*compliance with a legal obligation or in the exercise of controller's official authority*" 17(3)(b), and "*for archiving purposes in the public interest, scientific or historical research purposes*" 17(3)(d). Clearly, the exemptions described in the 17(3)(d) may as well refer, for many controllers, to the instances of their archived data. However, the GDPR Article 89, which provides "derogations relating to processing for archiving purposes in the public interest, scientific or historical research purposes or statistical purposes", does not include the RtbF (Article 17) as a potential exemption. What perplexes more the issue is the mismatch between the Article 89 and the Recital 156 as the later does provide for a derogation from the Article 17 when personal data are processed for the same purposes.[2] And although the recitals are not legally binding texts, yet the conditions under which an exemption from the RtbF is allowed are not crystal clear. Furthermore, as the Article 17(3)(b) clearly mandates, for certain cases other legal obligations stipulating data retention by the controllers will prevail over the GDPR's provisions.

As we discussed in 3.5.3, controllers need to implement technical solutions not only to allow the tracking of personal information but also to prove its efficient removal in the case of request for erasure under the GDPR. And although the first may not be considered a difficult task, since many controllers keep links of their copied

[2] *Member States should be authorized to provide, under specific conditions and subject to appropriate safeguards for data subjects, specifications and derogations with regard to the information requirements and rights to rectification, to erasure,* **to be forgotten**, ..."

information, the burden to prove that the erasure has been implemented successfully from all available sources is still technologically questionable. Taking further into account that the personal data may have been already backed up or archived by the controller or by the third parties, then the practical difficulty for implementing this requirement seems indisputable. As a matter of fact, implementing the RtbF requirement for personal data that have already been backed up or archived is deemed to be not an easy task.

Before proceeding with the study of the organizational, business, and technical implications of enforcing the GDPR in real case backup scenarios, we describe below what a backup process entails and the most prevailing international security standards specifying backup procedures.

4.3.2 The Process of Backing up

In general, backup is "*a copy of information held on a computer that is stored separately from the computer*".[3] Backups are considered to be fundamental processes within the business continuity plan as they allow for recovery when an information system suffers a disaster. The disaster may stem from various sources that include malicious actions, e.g. cyber attacks, but also physical damages, hardware failures and system crashes. Therefore, backups' goal is not data preservation, but quick recovery. Eventually, backup is a repeated process whose regularity depends on the criticality of the system and the data it stores. Considering the enterprises, it has already been shown that the data-loss phenomenon leads to serious financial loss in the scale of billions per year [13].

Practically, a backup is a copy of organization's data in a specific timeframe that can be recovered in case of a disaster. In fact, according to the dictionary of Storage Networking Industry Association (SNIA),[4] the Point In Time copy (PIT copy) is "*A fully usable copy of a defined collection of data that contains an image of the data as it appeared at a single instant in time. ... Implementations may restrict point in time copies to be read-only or may permit subsequent writes to the copy*". Therefore, each backup is stored in specific data formats, on specific mediums and is marked based on the employed system/software and the timestamp to trace the time instance it reflects. To guarantee its availability, the storage media undergo scheduled checks, and to verify its integrity, each backup is digitally signed and a log of these records is securely stored.

Although archives are often inseparably associated with the notion of backups, still backups and archives distinguish from each other in many aspects. While backups are primarily used for fast operational recoveries by taking periodic images of active data which are retained only for a few days or weeks, archives are typically designed

[3] https://dictionary.cambridge.org/dictionary/english/backup.

[4] http://www.snia.org/sites/default/files/SNIADictionaryV2015-1_0.pdf.

to provide ongoing rapid access to years of business information by storing versions of data that are no longer in use, not changing frequently and not required on a regular basis.

There are various types of backups that can be categorized by their content, their medium or their method. For instance, based on the content we can have simple copies of some files, database dumps, full system images, and snapshots. The choice of its content is normally subject to the restrictions an organization has on recovering for a specific system. Therefore, for highly critical systems that need to be instantly recovered, a snapshot of the system is stored and loaded when deemed necessary. The latter applies to cloud instances and to virtualized systems in general. Backups also vary in terms of the medium, as they may be stored in different mediums due to cost and durability constraints. Moreover, to reduce space requirements, we have incremental backups which contain only the data that have changed since the preceding backup, and differential backups which contain only the data that have changed since the previous full backup.

A well-known strategy for backups is the 3-2-1 rule. The core idea of this strategy is to minimize possible failures during the process of storing and recovering a backup. According to this rule, one must keep at least three backups. These three backups must be stored in two different mediums. From these three backups, one backup must be off-site.

4.3.3 IT Security Standards for Backup Procedures

In this section we present the most widely adopted international standards for IT security assurance, and especially those concerning backup procedures. Generally speaking, the standards set the primary requirements that lead to regulatory compliance and they are commonly categorized by region and/or by sector. For instance, the Health Insurance Portability and Accountability Act (HIPAA) [14] is a US regulation framework for ensuring the confidentiality and security of Protected Health Information (PHI) while the Payment Card Industry Data Security Standards (PCI-DSS) [15] is a financial industry standard aiming to protect payment card data used in transactions.

In the IT security domain, several standardization bodies, such as the International Standardization Organization (ISO) [16], the American National Standards Institute (ANSI) [17], the Canada's Standards Association [18] and the Standards Australia [19], have developed security and privacy frameworks that may be incorporated in organizations' processes and procedures to protect their data assets. These standards recommend also methodologies for IT governance, risk identification, security controls, and information security. More specifically, the ISO/IEC 38500 [20] introduces an IT governance framework that provides guidance in the case of cloud services. ISACA organization created a methodology named COBIT (Control Objectives for Information and Related Technology) for better information management and IT governance [21]. The National Institute of Standards and Technology (NIST) Cyber-security Framework (CSF) is a policy framework that focuses on the security of US businesses and private organizations against cyber attacks. Last but not least, the ISO/IEC 27000 series of standards [22] is probably the most widely known and used set of standards relating to the security of Information and Communication Technology (ICT) systems. In particular, ISO/IEC 27001:2013 [23] provides guidelines for mitigating risks of data breaches and fully supports the requirements of an information security management system, while ISO/IEC 27002 [24] provides best practice recommendations on information security controls. The ISO/IEC 27017 [25] is based on ISO/IEC 27002 and gives guidelines applicable to the provision and use of cloud services.

The compliance of an organization with the specifications of a standard is formally accomplished and demonstrated through specialized audits and respective certifications. As seen in Fig. 4.1, third party audits and certifications are most of the times set as prerequisites by vendors to ensure that the auditee is in compliance with regulatory mandates. Third party audits are conducted by independent bodies to verify that an organization conforms to the requirements of a chosen standard and continues to meet these requirements on an ongoing basis.

Backup compliance requirements are not referring just to the storage process of the data lifecycle. Instead, they cover a set of procedures that should be followed for keeping data assets securely and efficiently and, along with the backup restoration requirements, participate substantially in the Disaster Recovery and Business

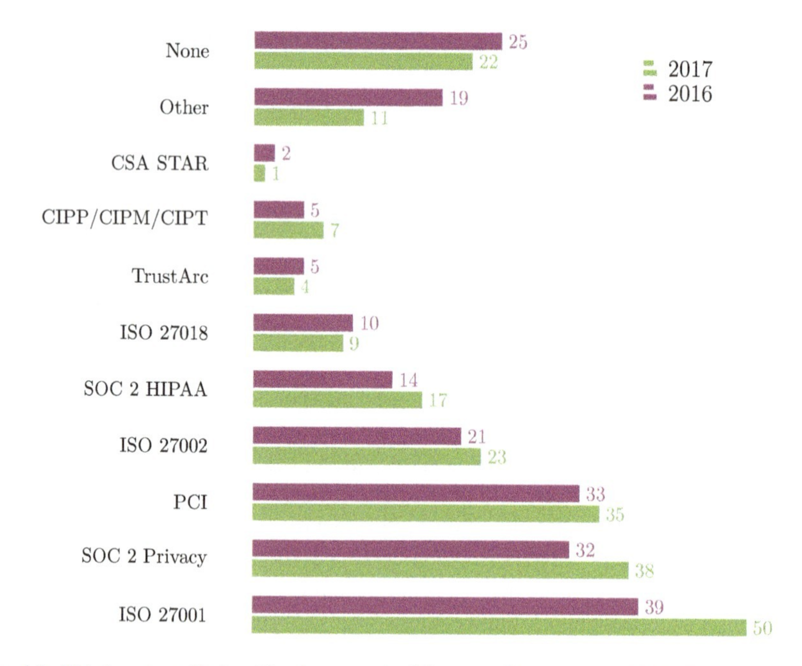

Fig. 4.1 Third party audits/certifications required from vendors. *Source* IAPP-EY annual privacy governance report 2017 [26]

Continuity plans. The appropriate policies for backing up personal data require an initial data mapping to be performed which should be then followed by an effective data governance model specifying how these personal data are to be managed in backups. In particular, the type of data that needs to be protected as well as the associated risks and privacy impact in case of a data breach must be clearly defined. On top, the methods and the means the data are stored and backed up as well as the relevant locations, including any off-site and on-site storage options, and the access to them should be specified. After mapping the data and establishing the appropriate data governance model, assurance that backups are well protected, and in special cases encrypted, should be provided. Furthermore, suitable measures ensuring that backups are kept only for a specified time need to be taken.

Likewise to security, backup policies and procedures are subject to the various requirements stemmed from the legal and regulatory frameworks. Although backup compliance is difficult to be achieved due to the growing number of both legal and regulatory requirements, the appropriate frameworks and standards selected for backup and recovery compliance navigate pertinently through the requirements that need to be met. Given the importance of data in every organization's operations, backup procedures are addressed by almost all security standards and frameworks, including COBIT [21], NIST [27] and Cloud Security Alliance (CSA) Cloud Controls Matrix (CCM) [28]. Under the ISO/IEC 27000 series [22] the backup procedures are covered by the ISO 27001:2013, (12 operations security, 12.3 Backups) in conjunction with the ISO 27040:2015 [29] which provides security and data protection guidance for storage systems. Additionally, the ISO/IEC 27018 standard [30], which is based on both ISO/IEC 27001 and ISO/IEC 27002 standards, provides additional guidance for the personally identifiable information (PII) stored and processed in public clouds and addresses storage and backup procedures taking into account the privacy principles of the ISO/IEC 29100 standard—Privacy Framework [16]. Relevant also to the backup processes is the NIST Special Publication (SP) 800-88 which, in conjunction with NIST SP 800-53, provides guidelines on sanitizing data storage media. Furthermore, in the case of terminating a cloud service used for maintaining backups, a well-defined and documented exit process is described in the CSCC document Practical Guide to Cloud Service Agreements [31].

4.3.4 Impact Analysis of Implementing the RtbF on Backups

As already mentioned, a major issue arising from the obligation for erasure requests under the GDPR RtbF concerns the case where personal data have already been backed up or archived. The issue occupies increasingly the interest of the IT industry since any non compliance may cause high sanctions. To this respect, technical experts debate on whether the RtbF should apply to the backups in the first place.[5],[6] Taken into

[5] https://www.linkedin.com/pulse/gdpr-right-forgotten-backups-jan-garefelt/.

[6] http://www.itgovernance.eu/blog/en/the-gdpr-how-the-right-to-be-forgotten-affects-backups.

account the enormous cost and effort of implementing the RtbF into the real-world backup and archive data stores, they argue that the most convenient interpretation of the RtbF to backups or archives would be not to be applicable at all. Yet, this view is not followed by most legal experts due to the absence from the GDPR text of any definite relevant exemption. As a matter of fact, the regulation does not provide any clear and unambiguous definition of the RtbF regarding its non-trivial practicalities of enforcing such a deletion when secondary uses apply, i.e. personal data have been disseminated to third parties, they have been anonymized or pseudonymized or they have been backed up and archived.

As both sides have valid arguments, the issue of how the RtbF is to be implemented on real IT systems is expected to be clarified upon an explanatory interpretation by an official EU data protection body such as the European Data Protection Board (EDPB), which replaced the WP29, or the European Data Protection Supervisor (EDPS).[7] Nevertheless, as up to now there is neither a clear basis nor an explicit derogation for backups within the GDPR text, it is reasonable to argue that the RtbF is indeed applied to backups as well. In fact, although the UK Data Protection Authority has already provided guidance arguing that deletion should also apply to the backups,[8] it recognizes in the meantime that deleting information from a system is not always a straightforward matter and hence sometimes it is preferable to put information “beyond use” providing appropriate safeguards for these cases.[9]

Notwithstanding this debate, the majority of the IT community agrees that the impact of the RtbF on the short-term backups, which are normally retained for a limited period of time to ensure business prompt recovery from accidental destroy or corruption, will be minimal comparing to the long-term archival backups that represent the long-term storage of the organization’s history records and they are used for future reference. Acknowledging this disproportionality, and for the sake of simplicity, we clarify that when the term backup is used hereafter it refers to the case of the long-term archival backup.

To overcome the oxymoron of having data deleted from archives while they have taken in the first place to safeguard the exact image of the data at a specific point in time, several solutions have been proposed. These include cryptographic erasures in which every record in a database is encrypted upfront with a different encryption key and upon a removal request the relevant encryption keys are deleted.[10],[11] Nevertheless, this method actually deactivates the personal data in question, rather than fullsy remove them. A second solution proposed by several analysts is to keep a separate table of all forgotten user IDs from the “live” system and each time a

[7] The European Data Protection Supervisor (EDPS) is an independent supervisory authority responsible for advising EU institutions on privacy related policies and legislation.

[8] https://ico.org.uk/for-organisations/guide-to-data-protection/guide-to-the-general-data-protection-regulation-gdpr/individual-rights/right-to-erasure/.

[9] https://ico.org.uk/media/for-organisations/documents/1475/deleting_personal_data.pdf.

[10] http://medium.com/@brentrobinson5/crypto-shredding-how-it-can-solve-modern-data-retention-challenges-da874b01745b.

[11] https://info.townsendsecurity.com/gdpr-right-erasure-encryption-key-management.

backup is restored, the forgotten users are checked against its contents and they are being re-forgotten.[12] Although this method seems a convenient workaround as it does not deal at all with the backups, it is questionable if it is appropriate to fulfill the GDPR requirements given the fact that the IDs in the separate table still constitute personal data, since with the use of additional information they can single out a specific person. Hence, the problem of forgetting these IDs from backups remains. What's more, this solution does not deal with the case where a person requires to remove only specific pieces of her personal information instead of all of it or when a portion of her information needs to be retained according to other legal requirements. Furthermore, none of the proposed approaches deals with the onerous matter of unstructured personal data, such as emails or files, when they need to be removed from backups.

Utterly, the implementation of the RtbF in the digital environment is not a simple task while its effective enforcement in the backups may be proved burdensome or even impossible in a number of scenarios. To emphasize the potential issues we identify and analyse below some affected areas in the organizational, business and technical domain.

4.3.4.1 Implications for the Standards

The RtbF that individuals may exercise under the GDPR involves requests for erasure of all or part of their personal data. As mentioned earlier, when these requests are received by a data controller the relevant data have to be removed under a specific timeframe from all the sites they reside, including archives and backups. Although the GDPR allows for exemptions from the RtbF when other legal obligations enforce the retention of these data, it is expected that the well established international standards driving and specifying backup procedures will most likely be questioned and challenged under the new law. This is due to the fact that the concept of backup specified by these standards mandate the storing of exact copies of the data as a fall-back mechanism that organizations should use only when things "go wrong", e.g. there is a physical medium failure, a disaster or a cyber-attack. Hence, the standards consider the backups to be immutable and thereby they are specifying that, apart from disruptions, the backup media should only be tampered with to check their health status.

Examples of standards that would be affected from the GDPR erasure requests are spread throughout the domains. In the US, HIPAA CFR 164.308 (7) (ii) (A) mandates:

> Data backup plan (Required). Establish and implement procedures to create and maintain retrievable exact copies of electronic protected health information

This may impact severely products and services compliant with HIPAA standards such as the Service Organization Control (SOC) 2 HIPAA.

[12] https://techblog.bozho.net/gdpr-practical-guide-developers/.

According to the ISO/IEC 27001 and 27002 Sect. 12.3.1, which specify that:

> Adequate backup facilities should be provided to ensure that all essential information and software can be recovered following a disaster or media failure.

one may rationally deduce that the backups cannot or should not be edited, as each data modification would not only affect the data but also the entire backup which by definition represents a unique instance at a given timestamp. The above concerns are also applied to ISO 27018 which adopts the backup procedure of 27002 with minor sector-specific guidance. Similarly, ISO 27040, Sect. 7.4.1., states that "*archival storage assumes a write-once, read-maybe access pattern, thus the integrity of the data in the system should be actively checked at regular intervals rather than waiting to when it is read*", indicating the immutability property of the archived information which is only read but not overwritten nor deleted.

In the case of PCI-DSS, the Sect. 10.5 of version 3.2[13] requirements reads as follows:

> 10.5 Secure audit trails so they cannot be altered.
> 10.5.1 Limit viewing of audit trails to those with a job-related need.
> 10.5.2 Protect audit trail files from unauthorized modifications.
> 10.5.3 Promptly back up audit trail files to a centralized log server or media that is difficult to alter.
> 10.5.4 Write logs for external-facing technologies onto a secure, centralized, internal log server or media device.
> 10.5.5 Use file-integrity monitoring or change-detection software on logs to ensure that existing log data cannot be changed without generating alerts (although new data being added should not cause an alert).

Moreover, the standard mandates in 10.7 to:

> Retain audit trail history for at least one year, with a minimum of three months immediately available for analysis (for example, online, archived, or restorable from backup).

While the standard in the above sections focus more on attacks from outsiders, it mandates that audit trails must not be altered and, regardless of other changes, the data must remain, even in the backups, for at least a year.

Plausibly, all the above standards, which according to Fig. 4.1 are among the most required standards in the market, present serious inconsistencies with the GDPR RtbF requirement and may result in difficult situations when organizations have to strike a balance between the regulation and the already widely spread well-known standards and best practices. While the compliance of the GDPR erasure obligations with the backup requirements mandated by the state-of-the-art security standards such as the ISO27000 series occupied recently the interest of both the research [32] and the security community,[14],[15] hitherto there has not been any comprehensive studies on this subject. Therefore, we argue that an immediate alignment of the standards

[13] https://www.pcisecuritystandards.org/documents/PCI_DSS_v3-2.pdf.

[14] http://www.iso27001security.com/ISO27k_GDPR_mapping_release_1.pdf.

[15] https://advisera.com/wp-content/uploads/sites/15/2018/03/List_of_documents_EU_GDPR_ISO_27001_Integrated_Documentation_Toolkit_EN.pdf.

with the GDPR provisions, and specifically with the RtbF, is deemed necessary and urgent. Otherwise, organizations, due to their high dependency on certifications and standards, may be severely affected.

4.3.4.2 Implications for the Data Retention Policies

While the GDPR does not mandate a specific timeframe for which personal data must be kept, data have actually a specific lifespan. In fact, data retention periods are determined by sector-specific business requirements and relevant domestic legislations. The storage limitation principle, according to which "*personal data shall be kept in a form which permits identification of data subjects for no longer than is necessary for the purposes for which the personal data are processed*", has been enshrined in the EU data protection law since the DPD era. Of course there are exceptions insofar as the personal data are processed solely for archiving purposes in the public interest, scientific or historical research purposes or statistical purposes (GDPR-Article 5(1)(e), DPD-Article 6(1)(e)). As a result, it can be safely reasoned that all contemporary systems that are processing, and thereby archiving, personal information are aligned with sector and domain specific retention requirements ensuring that data in the backups are not kept for more than it is necessary.

Yet, the GDPR obliges data controllers to ensure that the period for which the personal data are stored is limited to a strict minimum (Recital 39) and also to maintain a record, where possible, of their processing activities which shall include, among others, the envisaged time limits for erasure of the different categories of data (Article 30(1)(f)). Furthermore, the GDPR introduces stricter rules for facilitating the transparency of the process when data subjects are exercising their rights. To this end, it removes any fees relating to the administrative costs when controllers are being requested to remove any personal data under the RtbF, unless the request is "*manifestly unfounded or excessive*" (Article 12(5)), and specifies stricter timescales for responding to user requests for data erasure. More precisely, the GDPR enforces controllers to proceed with the erasure request "*without undue delay and in any event* **within one month** *of receipt of the request*" (Article 12(3)). This period "*may be extended by two further months where necessary, taking into account the complexity and the number of the requests*".

Apparently, these new requirements may present some technical challenges mainly due to the fact that user data are not stored within a single system but they are spread across multiple applications and storages, off-site and on-site, and they may be found under various forms such as emails, files, database records etc. Worse still, these data may have been already archived. To complicate even further, user data may have been archived in multiple backup files originated from various applications where they are used in. On top, they may have been included in many copies of the same backup file since backups for the same data are taken in regular periods of time. Last, typically each backup file includes data from many users. All the above imply that controllers need to search, identify and remove, in an efficient and timely manner within both the production and the backup environments, any relevant personal data an individual requested to be forgotten.

By exploiting economies of scale, many companies outsource their storage management, avoiding thus the costs of maintaining a data center. This shift has also been applied to backups as both cloud storage and backups are based on the same concept and they are performed usually by the same service providers. In terms of security, the general concerns about cloud backups are more or less the same as with the cloud storage. Therefore, cloud backups must ensure proper encryption of data, at least during transfer, while simultaneously it is essential to know where the backups are located in terms of geographical area as this, due to the diversity of data protection legislation across countries and continents, may entail issues with jurisdictions and fair information practices. Nevertheless, in the cases of cloud services, the providers do not know where the data of individual users reside, as by definition a cloud provider is agnostic of the data that stores.

Consequently, and regardless of whether the backups are performed on the cloud or not, it is of utter importance for data controller to keep track of the contents of each backup so that to either erase later by himself or to request from the service provider to remove any requested information. Finally, it should be noted that for the personal data to be entirely removed, they should not be simply deleted from the backups, but they have to be wiped out as well.

4.3.4.3 Implications for the Mediums

Irrespective of the followed security standards, the common practice for backup procedures are to oblige organizations to keep backups in disks—which may vary from optical CDs, DVDs to even bluetay discs and hard discs—or in tapes. While the cost per gigabyte for large capacity disk drives is constantly decreasing, tape backups are still cheaper. Regardless of whether full or incremental backups are performed by an organization, it follows that when data are needed to be removed from one copy, all the subsequent copies must be altered accordingly.

Nevertheless, while the digital records used and stored within the production systems can be easily removed once they are located, the same does not apply for the already backed up or archived data for which big effort, and hence further cost, is required. For instance, in the case of optical discs the data cannot be erased at all. As a result, when a user requests removal of her data, new copies of all the subsequent backup discs since user’s first data storage have to be made with the requested data omitted and the corresponding discs destroyed. Even in the cases of hard discs and tapes where there is no need to destroy the actual storage medium, not only the relevant data have to be erased but additionally all the following backups need to be appropriately altered.

Tampering with the backups is by no means easy as the stored data might be in a deprecated format, a fact that requires additional effort for efficiently searching through its contents, whereas the resulting new backups need to be properly signed and filed to ensure accountability in case of errors or undesired changes. It should be highlighted that while for some export formats deleting single records from the backup is typically allowed without needing to restore the full database, for tape backups this is impossible as tapes store data in sequential blocks and therefore cannot be randomly accessed. Therefore, deleting a record from a database table that resides in a tape backup implies the restoration of the whole database, thus increased cost and complexity.

4.3.4.4 Implications for the Search Services

Even when deleting single records from backups is typically allowed in order to conform to the exercise of the RtbF by individuals, searching into vast backup archives for particular personal data is definitely not an easy task. In reality, searching into backups for particular files is not a relative new feature since many pertinent research inventions have been so far patented [33–35] and several business tools[16],[17],[18] are already offering similar services. Recently, more sophisticated backup indexing and searching tools have emerged, such as the one provided by the Dell EMC Data Protection (DP) Search[19] which introduces unified index, search, and recovery features allowing easily backup search via a Google-like keyword search.[20] Nevertheless, none of these tools are scalable enough, especially when they need to search almost rapidly into massive archived storages of mostly unstructured data, which is still the most common type of data in every organization.[21],[22] Apparently, current technology seems to fall behind in methods for efficient search algorithms capable to look across the entire data landscape in a cross-platform and a cross-format manner without any noticeable delays. As a result, for effectively implementing GDPR-compliant backup and archiving search services, the technological limits of data processing are clearly required to be expanded.

4.3.4.5 Implications for ERPs and Analytics

Over the last decades, organizations worldwide have adopted Enterprise Resource Planning (ERP) software in order to automate and manage their business processes. Some of the most common ERP systems incorporate modules for product planning,

[16] https://support.code42.com/Administrator/5/Monitoring_and_managing/File_search/01_Enable_file_search_in_your_Code42_environment.

[17] https://docs.druva.com/001_inSync_Cloud/Cloud/030_Governance_DLP/030_Governance_and_DLP/010_Governance/Federated_Search_for_backed_up_data.

[18] https://helpcenter.veeam.com/archive/backup/95/em/searching_vm_backups.html.

[19] https://cpsdocs.dellemc.com/bundle/P_DP_PG/page/GUID-67803015-B5B6-4D19-90C0-91311D876CA7.html.

[20] https://blog.dellemc.com/en-us/make-it-rain-with-your-emc-hybrid-cloud/.

[21] http://breakthroughanalysis.com/2008/08/01/unstructured-data-and-the-80-percent-rule/.

[22] https://corporate.delltechnologies.com/en-us/newsroom/announcements/2012/12/20121211-01.htm.

purchases, supply chain, procurement, inventory control, product distribution, human resources, accounting, marketing and finance. Integrating these modules into a single system is considered as a prerequisite for an ERP system whose actual potential lies in using the data for analytics, data-driven business decisions, risk reduction, fast-track reporting and performance management.

Initially, ERP systems targeted more on the back-office software leaving the front-office functions to be dealt with cooperating software such as Customer Relationship Management (CRM) systems that communicated with customers in a more "direct" way. Nevertheless, modern ERP solutions integrate front office components, including even software solutions for mobile devices. Moreover, present-day ERPs also provide enterprises with functionalities to collaborate with their peers, realizing system-to-system interaction and data exchange.

Within a business workflow there is a variety of data streams that include personal or sensitive user information which is subsequently backed up. These data streams originate from fundamental software modules, such as the Business Intelligence (BI) module, the CRM module, the Front-Desk module, email interfaces and, as illustrated in Fig. 4.2, other modalities of user interaction, such as mobile phones and web pages.

Apparently, a software solution that collects, stores and analyses data of personal nature, e.g. related to customers, needs to oblige to data protection laws and hence to the GDPR RtbF provision. By examining primarily the storage and consequently the

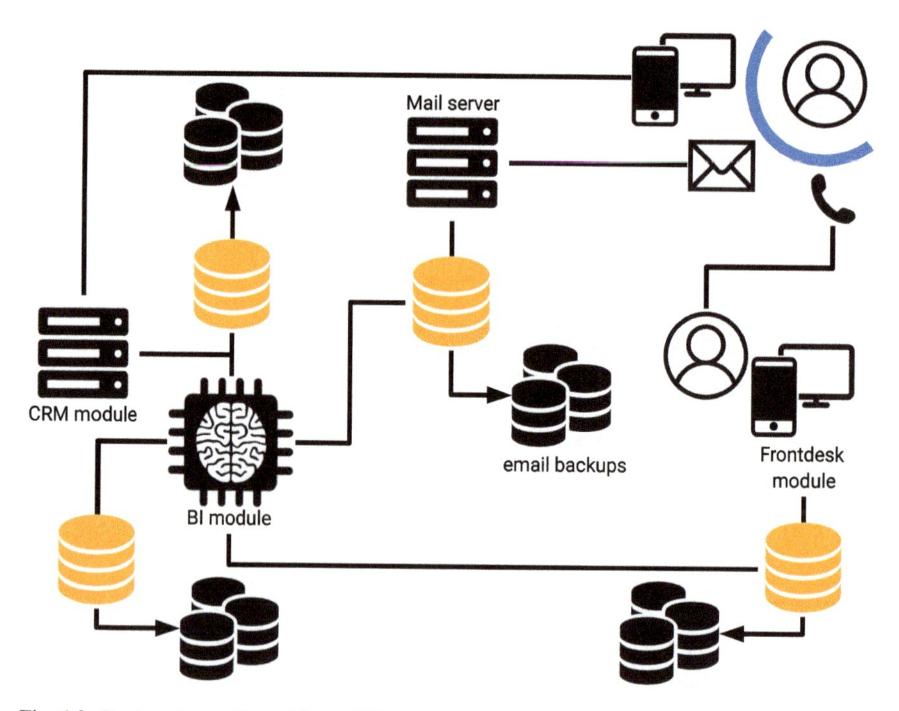

Fig. 4.2 Backup data collected from different streams within business workflows

backups of the involved personal data, perhaps the two most profound questions are whether it is possible to locate and erase personal data, when required, in a specified timeframe and whether these operations will hold back or even disable some of the ERPs' back-end functionality.

Locating the Social Security numbers or other key personal data in tens or even hundreds of thousands of distributed tables of e.g. a SAP ERP system (one of the most widely used ERPs to date) is not trivial[23] as this information is not always directly linked with the user ID of a single database. Personal data discovery on systems of such magnitude and complexity could require enormous amount of time and effort. Taken also into consideration that the GDPR, as we mentioned earlier, enforces specific and strict time constraints for controllers to respond to a user request for either deleting or accessing her personal data, the tasks of locating and removing specific piece of personal information from ERPs may face several challenges in the days to come. These challenges can be further intensified due to the fact that the backup plans of ERP installations may significantly vary both in terms of the means used (e.g. cloud infrastructures, hard copies), and the actual backup operations that range from daily database tables' copies to full ERP system image backups.

State-of-the-art ERPs already provide their enterprise customers with tools that can facilitate GDPR compliance in specific domains. For instance, SAP offers five tools to help address GDPR needs, namely the SAP Information Lifecycle Management [36], the SAP Data Services [37], the Information Steward [38], the SAP Process Control [39], and the SAP Access Control [40]. Yet, there are more tools for GDPR compliance available which can be categorized according to their key functionalities as follows:

- Tools that enable ERPs to discover where personal data are located in their systems.[24,25,26,27] While these tools can locate personal data residing in current systems, it is not clear whether they are able to locate personal data that have already been backed up. In order to achieve this, these tools should incorporate special logging mechanisms keeping track of backup data in their real-time databases.
- Tools that enable the deletion of personal data.[28] Since locating data does not necessarily mean deleting them, these functions are being investigated completely separately. This kind of tools is necessary to enable the "safe" deletion of personal data both for security and stability reasons since, apart from a secure data removal, there is a need for a guaranteed "stability" of the ERPs following the required deletion operations. As these tools can affect the integrity of the

[23] https://www.silwoodtechnology.com/blog/tested-5-erp-and-crm-packages-evaluated-for-gdpr-personal-data/.

[24] https://www.paconsulting.com/services/cyber-security-and-digital-trust/cyber-security/ediscovery/.

[25] https://www.silwoodtechnology.com/safyr/safyr-7-supporting-gdpr/.

[26] http://filefacets.com/.

[27] https://www.netgovern.com/ig-solutions/electronic-discovery.

[28] https://www.trustarc.com/products/individual-rights-manager/.

data, plausible challenges include the safekeeping of system's integrity when the removal of data both from backup and production environments is requested, as well as the successfully erasure of data that have been communicated with other parties (and possibly also backed up).
- Tools for managing and auditing the access to the stored personal data. These tools provide rules for reading and changing data files and hence they are essential towards the application of the new regulation.[29] Nevertheless, these tools refer to the run-time instances of the data and therefore it is not clear how these rules for managing access to personal data are going to be applied to the backup versions of the data.
- Tools for masking personal data. As already mentioned, ERP systems are able to integrate a large number of complex functions that include analytics, reporting and business decisions. Therefore, the elimination of a smaller or bigger part of a past database may affect the data analytics already produced or are to be produced in the future. To this end, an alternative to data deletion could be considered the masking of personal data. If this is deemed as the desired option for a use case, given the fact that the GDPR foresees for cases where the removal of personal data is indeed impossible *taking into account the available technology and the cost of implementation*, then the masking of all corresponding backed up personal data must also be ensured. However, many tools for masking personal data do not apply such changes to the production systems, let alone backups,[30] as the results may break the integrity of the system.

4.3.4.6 Archived ERP Data Use Cases

To thoroughly study the extent the GDPR regulation, and in particular the RtbF, affects the already established business operations, including backups, we illustrate below a number of use cases evolving from real case ERPs scenarios. More specifically, in what follows we present the sets of data fields that are stored and consequently backed up for a given customer under a well-known ERP installation for the telecommunications industry.

When a retail sale is performed, with invoice included, the ERP first checks if the corresponding customer exists in the system via a unique key field (e.g. Tax Identification/Registration Number, TIN/TRN). Following each user transaction, personal data relevant to the specific transaction are saved. An indicative list of personal data stored in such cases is illustrated in Table 4.1.

During dunning a significant set of personal data are collected, stored and archived. Dunning involves the process where a company communicates with its customers in order to insure the payment of their due amounts, a very common process supported by the ERP functionality. The communication is usually achieved through SMS

[29] https://www.trustarc.com/products/individual-rights-manager/.

[30] https://www.epiuselabs.com/data-secure.

Table 4.1 Archived objects from a retail sale

New user	User transaction
Name	Date of transaction
Surname	Time of transaction
Billing address	User who performed the transaction
Shipping address	The place the transaction took place
TIN/TRN	The method of payment
Method of payment	The payment terms (e.g. Installments, cash on delivery)
Bank information	The billing address and shipping address of that particular transaction
	The payment card number (encrypted) with which the payment may has been made
	Items bought in the transaction
	Quantity bought
	Price
	Discounts

messages, calls from representatives, and emails. The data objects that can be archived from the dunning process include:

- Phone Calls made to the customer.
- SMS messages sent to the customer.
- Legal offices the customer was assigned to.
- All previous dunning categories the particular customer has been in.
- Possible disconnections from the Subscription.
- Legal actions taken against the subscriber.

Finally, for every subscriber, regardless of whether she is active or inactive, the data depicted in Table 4.2 are archived.

Taking the above use cases into consideration one may reasonably deduce that the referenced personal and financial data are necessary for delivering telecommunication services while complying with tax and financial laws, and therefore their retention is not only justified but also mandatory. Any request under the RtbF for removing such data it will most probably collide with tax and financial legislations that oblige their collection, storing, and hence archiving, for a maximum retention period defined under domestic laws. Nevertheless, when this retention period expires businesses are now required to remove these data, from both the production and the backup environment, upon an individual's request for erasure under the RtbF. Notwithstanding this obligation, keeping financial data beyond the predetermined retention period for use in advanced data analytics and automated business decisions provides undeniably a valuable resource for supporting businesses' underlying operations, albeit particularly harmful in terms of profiling people's lives as we discuss later in Sect. 6.2.5. As a result, new challenges arise for corporations that have to find

Table 4.2 Archived objects related to a subscriber

Fields	
Name	Payment dates
Surname	Payment amounts
Sex	Payment methods
Address	Amounts owed
Telephone numbers he has used/using	TIN/TRN
Disconnection dates	Disconnection reasons
Equipment issued to customer (Routers, SetTopBox, etc.)	Serial numbers of equipment leased or sold
Service subscription history (e.g. Internet, cloud services, etc.)	Messages issued from company to subscriber
Materials bought	ID number
Call history (Company to subscriber and vice versa)	Passport number

technical alternatives of exploiting their valuable data stores while not compromising customer's data protection rights such as the RtbF.

All in all, the above analysis demonstrates that tampering with backups, regardless of whether it is intended or not, is neither a trivial task nor a straightforward process and it is heavily impacted by the data retention regulations, the mediums used for backups, as well as the relevant standards and the available technologies.

4.4 Towards GDPR Compliance

The concerns described in this Chapter are only some indicative examples of the long list of conflicts need to be resolved for the successful GDPR application. Beyond any doubt, the issues arising from the data protection provisions introduced by the regulation, and in particular the RtbF, are not minor. While there are some recent studies proposing technical solutions for a feasible implementation of the RtbF, most of them concentrate on the problem of implementing a RtbF compatible with the CJEU decision, that is a right to de-list personal information from search engines, and not the one enforced by the GDPR to entirely remove it (see Sect. 3.5.2). For example, in an analysis [41] of the RtbF in the context of the CJEU decision and its impact on search engines, an implementation approach based on Personal Data Management Architectures (PDMA) was suggested. However, the approach does not deal with the situation of permanently deleting information from all of the data controllers holding this information. On the contrary, in what follows, we focus on discussing indicative methods and frameworks that can be employed for implementing the permanent and parallel deletion of personal data upon request.

In this context, and in spite of earlier studies on establishing theoretical foundations for the design of mechanisms for forgetting of personal information [42], there exist numerous arguments against the feasibility of deleting information on the Internet, based mainly on the easiness of copying information, and hence, on the difficulty or impossibility of ensuring that information can be ever completely forgotten [43]. Indeed, while the industry has heretofore developed tools for facilitating users in their personal data administration, Novotny and Spiekermann in [44], after studying 13 available online services, concluded that, while half of them provide for some erasing mechanisms, none of them provides intelligent capabilities to forget outdated personal information. For example, Google has long ago introduced Google Dashboard[31] in the spirit of providing users with the capabilities of viewing, managing, and deleting their online personal information like web searches, shared docs etc. Still, the sheer amount of information the users view when they browse their dashboards usually have chilling effects about the extent of the collected information by Google who knows more about their Internet activity than they do.[32] While the deleting option for this information is a kind of relief, there is not any evidence of permanent erasure of these data from the company's servers. On top of that, one can never know if her data were replicated to other sites or services. Other industry efforts enabling the full control of personal data included personal data storages like TeamData[33] which used to enable users to securely manage data in their workplace by providing individuals with an online "data vault" where their work information was being stored or shared, while at the same time privacy by design principles were followed to assert privacy. Teamdata evolved now to another powerful service, digi.me,[34] which provides centralized management and control of one's all personal data and online accounts. Still, none of the platforms could assuredly remove permanently all of one's personal data.

As a matter of fact, while the Internet is spread with tech counselling articles and services on how one could delete all online personal information from various web services, there is not an automatic way to ensure the erasure of outdated or erroneous personal data from all of the services they may have been disseminated once they were uploaded. Even the "Web 2.0 Suicide Machine"[35] initiative launched in 2010, a tool that allows users to "suicide" their electronic selves in the social networks by automatically removing user's private content and friend relationships from these sites, operates with a handful of sites and does not guarantee full online removal. Given the futility of assuredly deleting online personal data, professional reputation managers implementing strategies that rely on techniques of burying offending information rather than removing it, have emerged recently. Nonetheless, these services address only the tip of the information iceberg [45]. In the light of this, and as more and more reputation queries are being processed by a handful of de facto reputation

[31] https://myaccount.google.com/dashboard.

[32] https://fossbytes.com/google-tracking-dashboard-myactivity/.

[33] https://teamdata.com/.

[34] https://digi.me/.

[35] http://suicidemachine.org/.

brokers, scholars have proposed a form of "reputation bankruptcy", a choice which will allow individuals to wipe their online reputation slates clean and start over after a predefined number of years.

The reputation bankruptcy idea evolves from the theoretical work for introducing forgetting in informational systems, suggested by Bannon in the early 21st century [46]. Bannon envisioned that private messages might be marked so that it is not possible to forward them without the author's permission, or that all social messages can be designed to fade away over time. He imagined various kinds of electronic tagging systems for messages that could time-stamp data and may contain something like a "sell-by" date in order to explore augmentation means for all human activities, both remembering and forgetting [46]. Similarly, Solove [47] imagined a world in which digital-storage devices could be programmed to delete photos or blog posts or other data that have reached their expiration dates, and he suggested that users could be prompted to select an expiration date before saving any data.

In this respect, Mayer-Schönberger [48] elaborated on the concept of forgetting through expiration dates for information. He described the various structural, legal, and technical components of expiration dates and how they would work together, while he offered a spectrum of possible implementations based on how thoroughly policy-makers and the public desire to revive forgetting [48]. Meanwhile, computer scientists have already initiated research on privacy-preserving ubiquitous computing frameworks and policies that enable enforcing limited retention periods for personal data storage [49–51] and they assure complete deletion of data and files [52, 53]. Following these lines, in [54] researchers demonstrated a relational database wherein once a tuple has been "expired", any and all its side-effects are removed, thereby eliminating all its traces rendering it unrecoverable, and also guaranteeing that the deletion itself is undetectable. Nevertheless, as it has been mentioned in many exploratory essays [1, 43, 55], the practicability of this theoretical principle is far from evident. Critics of the data expiration idea argue that even if auto-expire tools existed, they would do nothing to prevent the usual privacy problems when someone copies content from one site and moves it to another not supporting the auto-expire function.[36] This is the reason, as Mantelero notices in [56], why the idea of fixing a general time limit for mandatory erasure has been correctly avoided in the GDPR. Time, however, has been identified as a critical factor for introducing forgetting by many scholars, like Korenhof et al. [57] who argued that we should not overlook or disregard the importance of time in weighting the opposing interests when we are shaping policy mechanisms like the RtbF. Having this in mind, Korean researchers patented and sold a technique called Digital Aging System (DAS) which attaches "aging timer" to digital personal data [58], whereas the company Xpire[37] developed a smartphone app that enables the creation of self-destructing social posts in Facebook, Twitter, and Tumblr.

Acknowledging the importance of time and evolving the concept of data expiration, the idea of data degradation is also proposed. Privacy-aware data management

[36] http://blogs.harvard.edu/futureoftheinternet/2010/09/07/reputation-bankruptcy/.
[37] http://getxpire.com/xpireApp.

by means of data degradation, whereby sensitive data becomes less sensitive over time as a result of various degradation processes [59], is based on the assumption that long-lasting purposes can often be satisfied with a less accurate, and therefore less sensitive, version of the data. Data are progressively degraded such that they would still serve application purposes, even though their accuracy—and thus the privacy sensitivity—has been decreased. Yet, data degradation still faces the same weaknesses as those described for expiration dates since it cannot prevent the undesirable copy of data before their initial degradation.

There are also other theoretical approaches for achieving the forgetting of personal data which are based on the exercise of consent withdrawal. For instance, in the cases of biobanks it has been suggested [60] that the withdrawn samples and data to be parked in "limbo" or be dead-locked for a period of time and only destroyed/erased at the end of that period if the person withdrawing has not changed her mind. However, this approach cannot be legally accepted upon the GDPR's enforcement since the requirement for implementing consent withdrawal under the GDPR imposes data to be deleted "without undue delay" when an individual withdraws consent and the consent takes effect (in case neither of the exemptions described in Article 17(3) apply).

Pursuing the feasibility of forgetting in digital systems after a period of time, researchers, less than a decade ago, introduced Vanish [61], a very prominent technology for enforcing forgetting that causes sensitive information, such as emails, files, or text messages, to irreversibly self-destruct, hence "vanish", automatically after they are no longer useful, and all that without employing any centralized or trusted system. Vanish ensured that all copies of certain data become unreadable after a user-specified time even if an attacker obtains both a cached copy of the data and user's cryptographic keys and passwords. Instead of relying on data controllers to delete the data stored "in the cloud", Vanish encrypted the data and then "shattered" the encryption key. To read the data, the computer had to put the pieces of the key back together, but these "eroded" or "rusted" as time elapsed, and after a certain point the document couldn't longer be read. Vanish leveraged the services provided by decentralized, global-scale p2p infrastructures, and in particular by the Distributed Hash Tables (DHTs) for encrypting user data locally with a random encryption key not known to the user. The system destroys the local copy of the key and bits of the key are sprinkled across random nodes in the DHT. Moreover, it does so without any explicit action by the users or any party storing or archiving the data, in such a way that all copies of the data vanished simultaneously from all storage sites, online or offline. Unfortunately, while Vanish seemed a very promising solution, scientists managed to break it not long after its initial publication [62], and although the research team tried to tackle the problems identified in a following work [63], they never seemed to have succeeded their original goals. Yet, in the following years an increasing amount of encouraging follow up research works has been carried out for the quest of new improved versions of the Vanish prototype without facing its vulnerabilities [64–67].

Enforcing fine-grained data management obligations and improving the accountability of responsible parties as specified by policies and regulations is the focus of another research approach named Information Flow Control (IFC) [68, 69]. IFC is a data flow control model that enforces policy against every flow in the system. To achieve IFC, tags are linked with data and entities in order to represent various properties and policies concerning the data flow. The tags are collected into two labels: (a) a secrecy label representing the data's privacy/confidentiality/sensitivity; and (b) an integrity label representing the data quality/provenance/authority. IFC can assist with the erasure concerns coming under the GDPR's RtbF requirement as data flows are audited, and thus it is possible to determine where data has gone and to ensure that the deletion requests are directed to all relevant entities.

Extending the concept of IFC for managing personal data in the mobile environment, Enck et al. [70] proposed TaintDroid, an efficient, system-wide dynamic taint tracking and analysis system capable of simultaneously tracking multiple sources of sensitive data within the Android environment. TaintDroid provides real-time analysis by leveraging Android's virtualized execution environment to monitor the behaviour of third-party Android applications and to automatically label (taint) data from privacy-sensitive sources while transitively applying labels as sensitive data propagates through program variables, files, and interprocess messages. When tainted data leave the system, TaintDroid logs the data's labels, the application responsible for transmitting the data, and the data's destination. Although authors' primary goal was to detect when sensitive data leave the Android system, we believe that TaintDroid, along with other equivalent IFC models for cloud environments [68], could be used to provide visibility on how applications treat private data, and simultaneously, to satisfy forgetting requirements under the RtbF.

More recently, Zyskind et al. [71] proposed the use of the blockchain technology for the implementation of a platform that enables users to own and control their data without compromising security or limiting companies' and authorities' ability to provide personalized services. More specifically, they described and implemented a decentralized personal data management system and a protocol that turns a blockchain into an automated access-control manager that does not require trust in a third party. They accomplished this by combining a blockchain, re-purposed as an access-control moderator, with an off-chain storage solution and, at any given time, the user may alter the set of permissions and revoke access to previously collected data. Users are not required to trust any third-party, and they are always aware of the data being collected about them and how they are used. The decentralized nature of the blockchain, combined with digitally-signed transactions, ensures that an adversary cannot pose as the user or corrupt the network, since that would imply the adversary forged a digital-signature or gained control over the majority of the network's resources. Therefore, this decentralized platform makes legal and regulatory decisions about collecting, storing and sharing sensitive data much simpler because it is possible laws and regulations to be programmed into the blockchain itself, so that they are enforced automatically. In this respect, the proposed solution could be rendered as a very good candidate for implementing the RtbF requirement specified in the GDPR. Yet, taken into account our discussion on the privacy limitations of blockchains in Chap. 7, the adoption of such option is arguable.

Taking into account the data protection requirements enforced by the GDPR, Microsoft researchers [72] proposed an interoperable context-aware metadata-based architecture that allows permissions and policies to be bound to data, enabling this way any entity to handle the data in a way that is consistent with a user's wishes, including revoking a use previously granted. Within this architecture, a trusted processing container is used in order to ensure data are processed according to the policies specified by the associated metadata, even when they are temporarily separated for performance. When data leave the container, the metadata, which provide a layer of abstraction from the data, are reattached. Processing and interpretation would occur first on the metadata, whereas the data themselves can be used only when the entity has been properly validated as having the right to use the data. The architecture is flexible enough to allow for changing trust norms and to help balancing the tension between users and businesses. It also satisfies regulators' desire for increased transparency and greater accountability, while still allowing data to flow in ways that provide value to all participants in the ecosystem. Yet, while this metadata-based architecture is considered to be a useful building block for enabling and supporting the RtbF imposed by the GDPR, metadata alone cannot guarantee that entities will abide by specified policies. However, it can at least facilitate their enforcement by making them readily accessible, and, when implemented as part of a principles-based policy framework, it can enforce trustworthy data practices imposed by regulations.

References

1. I.S. Rubinstein, Big data: the end of privacy or a new beginning? Int. Data Privacy Law **3**(2), 74–87 (2013)
2. V. Kadenic, *Compliance of Data Lake Enterprise Architecture Model with the General Data Protection Regulation (GDPR)*. Bachelor thesis, Luleå University of Technology (2015)
3. M. Blanton, P. Gasti, Secure and efficient protocols for iris and fingerprint identification, in *Computer Security–ESORICS* (Springer, 2011), pp. 190–209
4. C. Blundo, E. De Cristofaro, P. Gasti, EsPRESSo: efficient privacy-preserving evaluation of sample set similarity, in *Data Privacy Management and Autonomous Spontaneous Security* (Springer, 2013) pp. 89–103
5. J. Bringer, M. Favre, H. Chabanne, A. Patey, Faster secure computation for biometric identification using filtering, in *2012 5th IAPR International Conference on Biometrics (ICB)* (IEEE, 2012), pp. 257–264
6. J. Bringer, H. Chabanne, A. Patey, Practical identification with encrypted biometric data using oblivious ram, in *2013 International Conference on Biometrics (ICB)* (IEEE, 2013), pp. 1–8
7. C. Patsakis, J. van Rest, M. Choraś, M. Bouroche, Privacy-preserving biometric authentication and matching via lattice-based encryption, in *International Workshop on Data Privacy Management* (Springer, 2015) pp 169–182
8. S.F. Shahandashti, R. Safavi-Naini, P. Ogunbona, Private fingerprint matching, in *Information Security and Privacy* (Springer, 2012), pp. 426–433
9. A.T.B. Jin, D.N.C. Ling, A. Goh, Biohashing: two factor authentication featuring fingerprint data and tokenised random number. Pattern Recogn. **37**(11), 2245–2255 (2004)

10. N. Ratha, J. Connell, R.M. Bolle, S. Chikkerur, Cancelable biometrics: a case study in fingerprints, in *18th International Conference on Pattern Recognition (ICPR'06)* (IEEE, 2006) vol 4, pp 370–373
11. A.B. Teoh, Y.W. Kuan, S. Lee, Cancellable biometrics and annotations on biohash. Pattern Recogn. **41**(6), 2034–2044 (2008)
12. F. Schaub, R. Balebako, A.L. Durity, L.F. Cranor, A design space for effective privacy notices, in *Eleventh Symposium On Usable Privacy and Security (SOUPS 2015)*, (USENIX Association, 2015), pp. 1–17
13. E. Kovacs, *Downtime and Data Loss Cost Enterprises* $1.7 Trillion Per Year: EMC (2014). https://www.securityweek.com/downtime-and-data-loss-cost-enterprises-17-trillion-year-emc
14. Health Information Privacy (2015). https://www.hhs.gov/hipaa
15. PCI Security Standards Council, *Download Data Security and Credit Card Security Standards* (2021). https://www.pcisecuritystandards.org/security_standards/
16. ISO—International Organization for Standardization, *Iso 29100 iso/iec 29100:2011-Information Technology—Security Techniques—Privacy Framework* (2011). https://www.iso.org/standard/45123.html
17. American National Standards Institute—ANSI (2021) https://www.ansi.org/
18. Canadian Standards Association, *Model Code for the Protection of Personal Information* (2013). https://www.scc.ca/en/standards/work-programs/csa/model-code-for-protection-personal-information
19. Standards Australia, *Personal Privacy Practices for the Electronic Tolling Industry; AS 4721-2000* (2000). https://www.standards.org.au/standards-catalogue/sa-snz/other/it-023/as--4721-2000
20. ISO 38500 (ISO38500) IT Governance Standard (2021). http://www.38500.org/
21. COBIT 5: A Business Framework for the Governance and Management of Enterprise IT (2021). http://www.isaca.org/COBIT/Pages/default.aspx
22. ISO—International Organization for Standardization, *An introduction to iso 27001, iso 27002....iso 27008* (2021). http://www.27000.org/
23. ISO—International Organization for Standardization, *ISO 27001 ISO/IEC 27001:2013-Information Technology—Security Techniques—Information Security Management Systems—Requirements* (2013a). https://www.iso.org/standard/54534.html
24. ISO—International Organization for Standardization, ISO 27002 ISO/IEC 27002:2013 Information Technology—Security Techniques—Code of Practice for Information Security Controls (2013b). https://www.iso.org/standard/54533.html
25. ISO—International Organization for Standardization, *ISO/IEC 27017:2015—Information Technology—Security Techniques—Code of Practice for Information Security Controls Based on ISO/IEC 27002 for Cloud Services* (2015a). http://www.iso.org/iso/catalogue_detail?csnumber=43757
26. IAPP-EY, IAPP-EY Annual Privacy Governance Report 2017 (2018). https://iapp.org/media/pdf/resource_center/IAPP-EY-Governance-Report-2017.pdf
27. Joint Task, Transformation initiative, security and privacy controls for federal information systems and organizations. NIST Spec. Publ. **800**(53), 8–13 (2013)
28. Cloud Security Alliance, *Cloud Controls Matrix* (2021). https://cloudsecurityalliance.org/group/cloud-controls-matrix/
29. ISO—International Organization for Standardization, ISO/IEC 27040:2015—Information Technology—Security Techniques—Storage Security (2015b). http://www.iso.org/iso/catalogue_detail?csnumber=44404
30. ISO—International Organization for Standardization, *Iso/iec 27018:2014—Information Technology—Security Techniques—Code of Practice for Protection of Personally Identifiable Information (pii) in Public Clouds Acting as pii Processors* (2014). http://www.iso.org/iso/catalogue_detail.htm?csnumber=61498

31. Cloud Standards Customer Council (CSCC), *Practical Guide to Cloud Service Agreements Version 2.0* (2015). http://www.cloud-council.org/deliverables/CSCC-Practical-Guide-to-Cloud-Service-Agreements.pdf
32. C. Bartolini, G. Gheorghe, A. Giurgiu, M. Sabetzadeh, N. Sannier, Assessing IT security standards against the upcoming GDPR for cloud systems, in Proceedings of the Grande Region Security and Reliability Day (GRSRD) (2015), pp. 40–42
33. D. Lyons, E. Weiss, P. Cisler, P. McInerney, J. Hornkvist, Searching and restoring of backups. US Patent App. 11/760,588 (2008)
34. A.A. Nene, S.P. Velupula, M. Kumar, A.V. Dhumale, A.G. Das, Backup search agents for use with desktop search tools. US Patent 7,890,527 (2011)
35. Y.P. Tsaur, R.R. Stringham, S. Sethumadhavan, Method and apparatus for performing file-level restoration from a block-based backup file stored on a sequential storage device. US Patent 8,386,733 (2013)
36. SAP Information Lifecycle Management (2018c). https://www.sap.com/products/information-lifecycle-management.html
37. SAP Data Services (2018b). https://www.sap.com/products/data-services.html
38. SAP Information Steward (2018d). https://www.sap.com/products/data-profiling-steward.html
39. SAP Process Control (2018e). https://www.sap.com/products/internal-control.html
40. SAP Access Control (2018a). https://www.sap.com/products/access-control.html
41. K. O'Hara, N. Shadbolt, W. Hall, A Pragmatic Approach to the Right to be Forgotten (2016), URL https://eprints.soton.ac.uk/389777/
42. D. Barua, J. Kay, B. Kummerfeld, C. Paris, Theoretical foundations for user-controlled forgetting in scrutable long term user models, in *Proceedings of the 23rd Australian Computer-Human Interaction Conference* (ACM, 2011), pp. 40–49
43. D. Lindsay, *The "Right to be Forgotten" is Not Censorship* (2012). http://www.monash.edu/news/opinions/the-right-to-be-forgotten-is-not-censorship
44. A. Novotny, S. Spiekermann, Oblivion on the web: an inquiry of user needs and technologies, in *Twenty Second European Conference on Information Systems* (Tel Aviv, 2014)
45. J.A. Burkell, Remembering me: big data, individual identity, and the psychological necessity of forgetting. Ethics Inf. Technol. **18**(1), 17–23 (2016)
46. L.J. Bannon, Forgetting as a feature, not a bug: the duality of memory and implications for ubiquitous computing. CoDesign **2**(01), 3–15 (2006)
47. D.J. Solove, *The Future of Reputation: Gossip, Rumor, and Privacy on the Internet* (Yale University Press, 2007)
48. V. Mayer-Shönberger, *Delete: The Virtue of Forgetting in the Digital Age* (Princeton University Press, 2011)
49. P. Ashley, S. Hada, G. Karjoth, C. Powers, M. Schunter, Enterprise privacy authorization language (epal) (2003)
50. J.I. Hong, J.A. Landay, An architecture for privacy-sensitive ubiquitous computing, in: Proceedings of the 2nd International Conference on Mobile Systems, Applications, and Services (ACM, 2004) pp. 177–189
51. M. Langheinrich, A privacy awareness system for ubiquitous computing environments, in *International Conference on Ubiquitous Computing* (Springer, 2002), pp. 237–245
52. R. Perlman, File system design with assured delete, in *Third IEEE International Security in Storage Workshop, SISW'05* (IEEE, 2005), pp. 6–pp
53. Y. Tang, P.P. Lee, J.C. Lui, R. Perlman, Secure overlay cloud storage with access control and assured deletion. IEEE Trans. Dependable Secure Comput. **9**(6), 903–916 (2012)
54. S. Bajaj, R. Sion, Ficklebase: Looking into the future to erase the past, in *2013 IEEE 29th International Conference on Data Engineering (ICDE)* (IEEE, 2013), pp. 86–97
55. J. Ausloos, The right to be forgotten-worth remembering? Comput. Law Secur. Rev. **28**(2), 143–152 (2012)

56. A. Mantelero, The EU proposal for a general data protection regulation and the roots of the & #x201C;right to be forgotten. Comput. Law Secur. Rev. **29**(3), 229–235 (2013)
57. P. Korenhof, J. Ausloos, I. Szekely, M. Ambrose, G. Sartor, R. Leenes, Timing the right to be forgotten: a study into "time" as a factor in deciding about retention or erasure of data, in *Reforming European Data Protection Law* (Springer, 2015), pp. 171–201
58. H.J. Lee, J.H. Yun, H.S. Yoon, K.H. Lee, The right to be forgotten: standard on deleting the exposed personal information on the internet, in *Computer Science and Its Applications* (Springer, 2015), pp. 883–889
59. N. Anciaux, L. Bouganim, H. Van Heerde, P. Pucheral, P.M. Apers (2008) Data degradation: making private data less sensitive over time, in Proceedings of the 17th ACM Conference on Information and Knowledge Management (ACM, 2008), pp. 1401–1402
60. S. Holm, Withdrawing from research: a rethink in the context of research biobanks. Health Care Anal. **19**(3), 269 (2011)
61. R. Geambasu, T. Kohno, A.A. Levy, H.M. Levy, Vanish: increasing data privacy with self-destructing data. in *USENIX Security Symposium* (2009b), pp. 299–316
62. S. Wolchok, O.S. Hofmann, N. Heninger, E.W. Felten, J.A. Halderman, C.J. Rossbach, B. Waters, E. Witchel, Defeating vanish with low-cost sybil attacks against large DHTs, in *NDSS* (2010)
63. R. Geambasu, J. Falkner, P. Gardner, T. Kohno, A. Krishnamurthy, H.M. Levy, Experiences building security applications on DHTs (2009a)
64. G. Wang, F. Yue, Q. Liu, A secure self-destructing scheme for electronic data. J. Comput. Syst. Sci. **79**(2), 279–290 (2013)
65. J. Xiong, X. Liu, Z. Yao, J. Ma, Q. Li, K. Geng, P.S. Chen, A secure data self-destructing scheme in cloud computing. IEEE Trans. Cloud Comput. **2**(4), 448–458 (2014)
66. L. Zeng, Z. Shi, S. Xu, D. Feng, Safevanish: An improved data self-destruction for protecting data privacy. in *2010 IEEE Second International Conference on Cloud Computing Technology and Science (CloudCom)* (IEEE, 2010), pp. 521–528
67. L. Zeng, S. Chen, Q. Wei, D. Feng, *Sedas: A Self-Destructing Data System Based on Active Storage Framework, in APMRC* (IEEE, Digest, 2012), pp. 1–8
68. J. Bacon, D. Eyers, T.F.M. Pasquier, J. Singh, I. Papagiannis, P. Pietzuch, Information flow control for secure cloud computing. IEEE Trans. Netw. Serv. Manage. **11**(1), 76–89 (2014)
69. J. Singh, J. Powles, T. Pasquier, J. Bacon, Data flow management and compliance in cloud computing. IEEE Cloud Comput. **2**(4), 24–32 (2015)
70. W. Enck, P. Gilbert, S. Han, V. Tendulkar, B.G. Chun, L.P. Cox, J. Jung, P. McDaniel, A.N. Sheth, TaintDroid: an information-flow tracking system for realtime privacy monitoring on smartphones. ACM Trans. Comput. Syst. (TOCS) **32**(2), 5 (2014)
71. G. Zyskind, O. Nathan et al., Decentralizing privacy: Using blockchain to protect personal data, in *Security and Privacy Workshops (SPW)*. (IEEE, 2015), pp. 180–184
72. S. Maguire, J. Friedberg, M.H.C. Nguyen, P. Haynes, A metadata-based architecture for user-centered data accountability. Electron. Mark. **25**(2), 155–160 (2015)

Chapter 5
State-of-the-Art Technological Developments

Abstract In the previous chapter we analysed the impact of implementing the GDPR, and in particular the RtbF, in established IT environments and business processes. However, two advanced technological trends of our times used increasingly nowadays for storing and processing personal data, have been emerged in parallel and independently of the GDPR: the ubiquitous mobile computing and the decentralized p2p networks. The revolution of sensor capabilities embedded into the latest smartphones and the rapid evolution of machine learning algorithms have paved the way towards the recent advancements in mobile computing and sensing. At the same time, the re-introduction of decentralization nowadays is demonstrated by the boom of the blockchain and decentralized file storage and sharing networks such as the IPFS. Notwithstanding the unquestionable benefits that these state-of-the-art technologies bring to our daily lives and to our society, they simultaneously pose great risks to our privacy and data protection rights. Before proceeding in investigating the relevant privacy threats and how these technologies can comply with the GDPR principles in the next chapters, we present and examine hereafter the main characteristics and properties of the ubiquitous mobile computing and two commonly used decentralized p2p networks, the blockchain and the IPFS.

5.1 Introduction

In the previous Chapters, we analysed the impact of enforcing the GDPR, and in particular the RtbF, in established IT environments. However, there exist two advanced technological trends emerged in parallel and independently of the GDPR, whose increasing use for storing and managing personal data pose great uncertainties to data protection and privacy rights: mobile ubiquitous computing and sensing, and decentralized p2p networks.

The revolution of sensor capabilities embedded into the latest smartphones and the rapid evolution of machine learning algorithms have paved the way towards the recent developments in mobile computing and sensing. The emerging products of mobile sensing combined with the progress of the research in affective computing, a subfield of the Human Computer Interaction (HCI), have also resulted in great advances

E. Politou et al., *Privacy and Data Protection Challenges in the Distributed Era*,
Learning and Analytics in Intelligent Systems 26,
https://doi.org/10.1007/978-3-030-85443-0_5

in the area of social, health and psychological sciences and have provided powerful tools for analysing people's social behaviour and predicting their personalities and emotions.

At the same time, as mobile computing and the IoT extend the ways that data collection and computation are applied, we have entered into a period of rapid decentralization, characterized by increasingly powerful decentralized devices, distributed computing, ubiquitous network access, and decentralized storage resources.[1] As a matter of fact, decentralized computing is more relevant today than any other time in the history of information technology, given that in recent years decentralization has been re-introduced as a means to assure the reliability of non-trusted environments such as those of cryptocurrencies. The hype of decentralized computing nowadays is most successfully demonstrated by the boom of Distributed Ledger Technologies (DLTs) and decentralized file storage and sharing networks. The most well-known DLT thus far is the blockchain, while the most widely used decentralized file sharing platform is the InterPlanetary File System (IPFS).

Notwithstanding the unquestionable benefits these state-of-the-art technologies bring to our daily lives and to our society, they simultaneously pose great risks to our privacy and data protection rights. Before proceeding in investigating the relevant privacy threats and the alignment of these technologies with the GDPR principles in the next Chapters, we present and examine hereafter the main characteristics and properties of these two emerging technologies.

5.2 Mobile Ubiquitous Computing

Since the mid-1990s when IBM introduced "Simon", the first-ever smartphone [1], today's smartphones with multi-core CPUs and gigabytes of memory are placing more processing capabilities in individuals' pockets than computers of past decades placed on people's desktops [2]. Besides the impressive features of CPU and memory, modern smartphones are programmable devices equipped with a range of cheap though powerful embedded sensors, such as gyroscopes, GPSs, accelerometers or magnetometers, which enable the development of personal and community-scaled sensing applications [3]. This rise of rich-sensor smartphones has enabled the recent birth of mobile sensing, an emerging and exciting interdisciplinary research field that requires significant advances in mobile computing, machine learning and systems design [4]. There are several types of mobile sensing such as individual, participatory, opportunistic, crowd, social, and on top of that, the object of sensing can either be people-centered or environment-centered [5].

Meanwhile, ubiquitous computing has already changed significantly the way computing and communication resources are used nowadays by fostering the principles of connectivity with any device, in any location and in any format [6]. Supporting these principles, Weiser [7], ubiquitous computing pioneer who first coined the term

[1] https://samsungnext.com/whats-next/a-brief-history-of-decentralized-computing/.

ubiquitous, described the requirements to be met in order devices to be considered ubiquitous: to be cheap, low-power computers that include equally convenient displays; a network that ties them all together; and software systems implementing ubiquitous applications.

As a matter of fact, smartphones represent the first truly ubiquitous and pervasive mobile computing devices since their mobility and afforded computational power allow users to interface directly and continuously with them, more than ever before, while their embedded sensors open up smartphones to categories of applications which until recently were possible only through wearable sensors [4]. Wearable badges are ubiquitous devices which, just like smartphones, are utilizing a powerful set of sensors and utilities to monitor biometric signals or location data of their holders in order to provide healthcare interventions or customized driving directions respectively, and thus to assist users in their daily tasks. Yet, while wearable sensors are portable and promising, are still not viewed as personal companions. In contrast, sensor-enhanced smartphones always accompany users since they are willingly carried by a large fraction of people in developed countries and therefore constitute a rich information source [8, 9].

The volumes of multimodal data collected from people's daily use of smartphones through sources such as GPS, call logs and Bluetooth, enable the development of data collection tools to record various physical and behavioural aspects of users, ranging from how a device is used across different contexts to the analysis of spatial and social dimensions of users' everyday lives [10]. An early research study [11] showed that data collected from mobile phones could provide insight into the relational dynamics and behaviour patterns of individuals, whereas quite recently scientists demonstrated that physiological parameters such as heart and breathing rates could be recovered from a smartphone via accelerometer measurements while the person is carrying it in different locations or using it during different activities [12]. This pervasive, unobtrusive and cost-effective access to previously inaccessible sources of information on everyday social behaviour, such as physical proximity of people, phone usage and patterns of movement [8], offers new opportunities to researchers as it allows them not only to understand the impact of context on user behaviour but also to study individual differences, such as users personality and emotions. In turn, it can enable the design of communication features and multiple mobile applications that are tailored to the individual needs and preferences of a user [10].

Against this background, we explore below the field of mobile affective computing where data acquired through mobile devices are exploited by intelligent machine learning methods to infer people's emotions, traits and behaviours [13, 14].

5.2.1 Affective Computing

Affective computing is a subfield of HCI named after the field of psychology in which "affect" is basically a synonym for "emotion". The fundamental concepts of affect recognition, interpretation and representation were firstly introduced by

Rosalind Picard in 1997 [15] who understood the potential of computers to recognize, understand, express and reproduce human emotions. As Picard elaborated on many works, in the interaction between human and computers, a device has the ability to detect and appropriately respond to its user's emotions [16].

Indubitably, a computing device with today's capacity could gather cues to user emotion from a plethora of sources, such as facial expressions, posture, gestures, speech, the force or rhythm of keystrokes and the temperature changes of the hand on a mouse. All these sources could signify changes in users' emotional states and can effectively be detected, interpreted and correlated by a computer [17]. As a matter of fact, the recognition and detection of human affect are usually referred to with the generic term "emotion detection" or "emotion recognition" which actually denotes the task of recognizing and classifying a person's emotion, such as anger, happiness or stress, across all possible channels of communication (modalities). Emotion recognition leverages techniques from multiple areas, such as signal processing, machine learning, and computer vision and may apply to a wide area of sciences, from psychometry and sociology studies to marketing and surveillance applications. Ultimately, the research field of affective computing is about studying and developing systems and devices that are able to recognize, interpret, process, correlate and simulate human affects [18].

5.2.2 *Mobile Affective Computing and Ubiquitous Sensing*

Up to 10 years ago, the basic approach of affective recognition process was to observe a person's patterns of behaviour via sensors such as cameras, microphones or pressure sensors applied to objects the user is in contact with (mouse, chair, keyboard, steering wheel, toy) and use computers to associate these patterns with probable affective state information [19]. Nevertheless, the boom of mobile technology during the last decade has radically changed the way machines sense human emotions. Mobile devices do not only constitute just a telecommunication device in people's everyday lives but, due to the recent technological advances, they are gradually replacing computers in all aspects of the digitized world. This large shift in the world of personal computing indicates clearly that users are willing to sacrifice performance in the name of portability and price, as smartphones are cheaper (since their cost is often folded into the cost of a multi-year contract with a mobile services provider), lightweight and can fit in a purse or pocket. Still, high-end smartphone processors today are faster than PC processors from years ago, or very low-end PC processors today and, according to the latest statistics, mobile web browsing has already overtaken the web browsing in desktops.

The mobile evolution does not only outline an era in which powerful machine-learning algorithms for statistical inferences using sensor data can be designed to run on commodity phones but, first and foremost, it facilitates the monitoring and analysis of human behaviour and social interactions on a large scale and in nearly real-time [2, 9]. It is against this background that social scientists, having as an ultimate goal the individuals' wellbeing and sustainable living, are keen on locating and isolating the unique social and personal conditions that impact people's affective state and, vice versa, on discovering personality characteristics that affect social behaviour. Taking into account that people' personality has been found to influence their behaviour in social interactions, several recent studies have investigated personality traits and their relationship to the use of the Internet and the social media [20]. In the context of mobile computing, however, besides the potentials of discovering behaviour patterns of individuals from mobile sensing applications, smartphone data can also provide meaningful associations between mobile phone usage and affective states, such as personality traits and emotions of individuals.

Inferring, recognizing and processing people's affective information, such as emotions, moods and personality, based on features extracted from their smartphones has been the object of extensive research by computer scientists, information engineers, applied psychologists and other relevant disciplines and, on that ground, the associated literature is rich and diverse. Technologies and methods used for accomplishing the concerned tasks vary significantly both in terms of employed smartphone modalities as well as in terms of chosen machine learning algorithms to infer and classify affective statuses. On top of this, pilot applications built for demonstrating the potentials of each model vary hugely in regard to experimental settings and environment. For instance, in the research study described in [21] the scientists found that the use of web, music, video, maps and other applications together with the traditional call and SMS usage, the proximity information derived from Bluetooth, and the use of a camera, can be indicators of the Big-Five personality traits. Similarly, in [22] the use of anonymized usage data extracted from smartphone call records, including variables obtained from the social network analysis of the calls, are employed to automatically predict users' personality (as characterized by the Big Five model). In [23] the researchers demonstrated that user personality could be reliably inferred from basic information accessible from all commodity smartphones. The features used in the research fall under five broad categories: basic phone use (e.g., number of calls, number of texts), active user behaviours (e.g., number of calls initiated, time to answer a text), location (radius of gyration, number of places from which calls have been made), regularity (e.g., temporal calling routine, call and text inter-time) and diversity (call entropy, number of interactions by number of contacts ratio).

Many research studies also infer and classify users' emotions by inconspicuously collecting and analysing user-generated data from different types of smartphone sensors and utilities. For instance, in [24] a mobile sensing platform recognizes user's emotions by processing the outputs from the sensors of commodity smartphones. Accelerometer, Bluetooth and GPS, together with microphone inputs, are used to infer participants' emotional states of happy, sad, fear, anger and neutral, based on users' movements, location, proximity and conversation with other users. Features extracted from smartphone data such as typing speed, touch count and device's shake collected while users are using a certain application on their smartphones (i.e. writing tweets), as well as location and weather information are used in an Android application [25] to recognize users' six basic emotions of anger, disgust, fear, happiness, sadness and surprise, plus one neutral. Likewise, in [26] an emotion recognition framework is proposed that demonstrates how simple touching behaviours can be used for recognizing smartphone users' emotional states. The authors of [27] built a model for automatic recognition of daily happiness based on information obtained from mobile phone usage data (call logs, SMS and Bluetooth proximity data) as well as indicators coming from weather factors, whereas Pielot et al. in [28] demonstrated the detection of boredom by utilizing mobile phone usage.

MoodSense [29] gathers information already available in smartphones (SMS, email, phone call, application usage, web browsing and location) and demonstrates that user mood can be inferred and classified into four major types with a quite satisfactory accuracy, whereas MoodScope is a smartphone software system that statistically infers user's daily mood by analysing communication history and application usage patterns. Oriented towards the prediction of stress in working environments, researchers studied the use of only one smartphone built-in sensor, the accelerometer, in order to detect behaviour correlating with subject's stress levels [30]. Stress recognition can also be inferred more directly through cameras and microphones embedded in subjects' smartphones. StressSense [31] is a voice-based stress detection system for recognizing the cognitive load related stress in job interviews and outdoor job execution tasks. Under the work carried out in [32] a system for predicting the negative states of depression, anxiety and stress based on mobile phone usage patterns, including calls and application usage, was proposed.

The above are some indicative state-of-the-art studies that leverage smartphone data to predict people's affective states. The complete list of the most prevailing, novel and promising works in the area of emotion recognition and personality inference through smartphones are presented in our survey in [14]. For compiling the survey, 42 studies published during the 7-year period up to its publication have been taken into account. The studies concern the detection of emotions, moods, personalities or other behavioural characteristics, like wellbeing, based on smartphone derived data and they were categorized by their adopted affective model for classifying subject's emotions, moods, or personality. Following a thorough analysis on these emerged works, we classify them by the type of mobile usage or sensors employed to provide the necessary data for inference. We then explore the various ethical and technical challenges faced by the research in mobile affective computing.

5.3 Decentralized p2p Networks

The core concept of decentralization has long been applied to political science, law, public administration, economics, and technology to describe the distribution or delegation of power away from a central authority [33, 34]. In computing, it is generally understood as the transfer of authority and control from a single central entity, and a single point of failure, to numerous localized peers that provide a robust infrastructure. Nowadays, decentralization in computing is almost synonymous with blockchain technology and other p2p networks, like BitTorrent,[2] which allow the secure and reliable sharing of information across peers without a single entity controlling the network.

However, decentralization of information systems is not a new idea. Even from the early '60s distributed and decentralized architectures were introduced to eliminate the problem of a single point of failure and to increase systems' robustness and redundancy [35]. It is worth pointing out, however, that while the terms decentralized and distributed are commonly used interchangeably to denote the lack of a central point of control, they actually have a subtly different meaning; the former is used to describe the conceptual and logical model of control, while the latter describes the technical characteristics of the infrastructure used to be built upon [36]. While p2p systems such as Usenet[3] seemed to be the most natural approach to the early days of the computer revolution, the prevalence of less expensive and less functional desktop PCs in the late '80s resulted in the predominance of client-server architecture [37]. However, the widespread adoption over the past 20 years of content sharing applications and protocols such as Napster[4] and BitTorrent which allowed millions of users to connect directly and share content over the Internet in a free and censorship resistance manner, returned decentralized p2p networks to spotlight. Meanwhile, the prevalence of grid and distributed computing in the last few decades relied also heavily on p2p foundations to allow the use of spare computing resources to complete high-performance tasks [37].

Since the dawn of online social networking, decentralization has also been proposed as an alternative for enhanced privacy and personal sovereignty in online social context [38]. As of 2009, the boom of blockchain technology with the advent of bitcoin caused the emergence of a new wave of decentralized p2p systems. Blockchain is literally a distributed ledger stored on a network of machines and is formed by a chain of blocks connected to each other using hash codes. Bitcoin, the first application of blockchain technology to decentralize financial transactions by creating a "*peer-to-peer electronic cash system*" [39], leveraged and improved upon various notions of p2p computing, and introduced, in effect, decentralization as a means to assure the reliability of non-trusted environments such as those of electronic currencies, i.e. cryptocurrencies. Nowadays, cryptocurrencies are usually discussed in the context of blockchains and Distributed Ledger Technologies (DLTs), terms closely interrelated but not identical. In what follows, we summarize and clarify the notions of DLT, blockchain and cryptocurrency and highlight their respective differences.

DLT

A DLT is a distributed digital ledger stored on a network of machines. Any changes to the ledger are reflected simultaneously for all holders of the ledger while the information stored is authenticated by a cryptographic signature [40]. The decentralized nature of the DLT eliminates the need for a central authority or intermediary to pro-

[2] https://www.bittorrent.com.

[3] http://www.usenet.com/what-is-usenet.

[4] http://www.britannica.com/topic/Napster.

cess, validate or authenticate transactions. At their core, DLTs are data structures to record transactions and set of functions to manipulate them. While each DLT differentiates itself using different data model and technologies, generally all DLTs are based on three well-known technologies: public key cryptography; distributed p2p networks; and consensus mechanisms. All three are blended in a unique and novel way to operate in an untrusted decentralized environment [41].

Blockchains

Even though blockchain technology was first outlined in 1991 as an effort to implement a system where document timestamps could not be tampered with [42], it was not until January 2009 that blockchain attracted worldwide attention when its first real-world application, the bitcoin cryptocurrency, was launched [39]. Due to its radical breakthrough and disruptive impact on financial services, blockchain is commonly discussed in the context of decentralized computing in the field of cryptocurrencies. However, blockchain applications go beyond finance to healthcare, supply chain management, and identity management, among others, to distribute information securely, transparently, and immutably [43].

Although the terms DLT and blockchain are often used interchangeably in the literature, they are not equivalent. For instance, while a blockchain is a sequence of blocks, DLTs do not require such a chain. As a matter of fact, a blockchain is just one type of DLT formed by a linked list (chain) of blocks connected to each other using hash codes, where each block references the previous block in the chain. Each block may contain a series of transactions which can be data of any sort. In blockchains, the transaction data are continuously appended, and they can be accessed by all the network participants (nodes). Essentially, blockchains are distributed and append-only ledgers that store transactions history while they provide a set of features that differentiate them from the other DLTs: smart contracts, which are pieces of executable code residing on the blockchain and executed once specific conditions are met; and miners, which are mining new transactions into the blockchain and can benefit financially from these mining activities [41].

Cryptocurrencies

While there have been multiple attempts during the last 30 years to solve the complex issues surrounding digital currencies [44–46], this was not achieved before 2009 when the bitcoin was launched. Generally speaking, the term cryptocurrency refers to a decentralized cryptography-based currency. Cryptocurrencies can be seen as asset resources or tokens on a blockchain network, and they are just one of the many possible applications of blockchain. Arguably, the true value of blockchain technology goes far beyond cryptocurrencies, whereas a blockchain can stand on its own just fine - no cryptocurrency needed [47]. In fact, there are already blockchain frameworks without any built-in cryptocurrency [48]. Yet, cryptocurrencies currently underlie most of the public blockchain applications to facilitate and incentivize their transactions.

Although bitcoin is currently the dominant cryptocurrency used in decentralized payments, the number of alternative cryptocurrencies (altcoins) has already surpassed

5000 [49]. In the context of cryptocurrencies like bitcoin, their name refers to more than the token, i.e. the currency itself that is traded in transactions or exchanges. It also denotes the underlying technology used: the protocol and the software system used to transfer the token over the blockchain network, as well as the blockchain network itself [50]. Nevertheless, while referring to the token as the technology can be right in the case of bitcoin, this is not the case when dealing with other blockchain projects like ethereum[5] [51] where the technology is known as ethereum, the native token is ether, and transactions are paid in gas.

Due to blockchain's profound impact on modern decentralized p2p services, in what follows we discuss blockchain's evolution and we present its key characteristics. We then present the most prominent decentralized p2p systems for file storage and sharing and we delve into the IPFS which is used widely nowadays to store off-chain data in its decentralized p2p network.

5.3.1 Blockchain

Blockchain technology dominates today's news, discussions, and articles, whereas its initiatives proliferate across industry and academia. Yet, few technologies today are as misunderstood as the blockchain. For some, blockchain is just a hype, an immature solution [52], an exaggerated bubble [53], or even a crypto-medieval system [54]. For others, it is an undeniably ingenious invention, an advance, a revolutionary technology. Blockchain's technological breakthrough has been even compared to the one brought by the use of the TCP/IP to modern computing, or the one Linux brought to modern application development [50, 55–58]. In addition, the bitcoin, the first cryptocurrency that exploits blockchains, has been called as "digital gold" [59], while ethereum, the largest open-source blockchain-based distributed computing platform, has been characterized as the backbone of the new Internet [60].

According to Nakamoto's original bitcoin paper, the proposed system is "*a system for electronic transactions without relying on trust*" [39]. Undoubtedly, blockchain's substantial impact on current and future real-world applications is attributed to its most highly acclaimed quality, its trustlessness. Trustlessness stems from blockchains' inherent security and transparency which eliminate the need for third-party intermediation and consequently the need for trust among users in decentralized and untrusted environments [61]. Yet, blockchain is not an alternative to trust but it is a new form of trust that tries to remove dependence on specific entities [62]. Blockchain actually decentralizes trust from a single central entity to millions of users worldwide. A highly praised property that underpins the blockchain's secure and transparent nature, and therefore guarantees its transactional integrity and auditability, is immutability. Blockchain's immutability arises from the fact that transaction data residing in blockchains are tampered-proof, i.e. they can neither be removed

[5] http://ethereum.org.

nor mutated. This append-only data structure signifies the permanent storage and availability of the stored information to everyone in the blockchain network.

Even though its underlying technology existed long before Satoshi Nakamoto published his paper on bitcoin [39], the immense and profound impact the bitcoin had in financial trades worldwide revealed blockchain as a new highly promising direction for decentralized computing. In the wake of the 2008 financial crisis where consumers' trust in banking was shaken, bitcoin's notion of decentralized financial systems seemed particularly appealing. Nevertheless, while blockchain technology is commonly associated with bitcoin and other cryptocurrencies, these are just the forerunners of a whole new wave of blockchain applications. According to experts, apart from disrupting financial services, blockchain could end up transforming a number of important industries, from healthcare to politics, whereas it has the potential to create new foundations for economic and social systems [43, 50, 63]. As most of its broad possible applications are still emerging, the future orientation and impact of blockchain technology cannot be easily predicted.

Still, its first stages of development during its decadal lifetime are beyond any expectations: blockchain has rapidly moved from bitcoin to thousands of altcoins, from decentralized electronic payments to complex tokens governed by Decentralized Autonomous Organizations (DAOs), and from the fat protocols stage where all value is generated in the protocol layer to the fat Decentralized Applications (DApps) stage where transactions are programmable by smart contracts [50, 64]. As this new generation of applications is evolving, blockchain's technical characteristics and specifications are becoming even more advanced and sophisticated [43]. By all means, describing in detail all the blockchain's features and functions is beyond the scope of this chapter. Instead, for the sake of simplicity, in what follows we delve into blockchain's key characteristics that are relevant to our following discussion in Chap. 7 regarding blockchain's collision with privacy: permission models, consensus protocols used for maintaining trust, and most importantly, immutability.

5.3.1.1 Types of Blockchains

Typically, blockchains are categorized based on their permission model, which determines the permissions that are granted to participants of a blockchain network, as well as based on their openness which determines who can participate in their network. As such, blockchains can be either permissionless or permissioned and also can be divided into public or private. The latter categorization is generally used to describe whether a blockchain is open to literally anyone with an Internet connection. In a public blockchain, anyone can join whereas a private blockchain operates under the authority and control of one or more organizations and only nodes from this organization are allowed to participate. In cases of more than one organization participating in the blockchain, then it is called consortium, or otherwise federated, blockchain.

As far as the permission model is concerned, in a permissionless blockchain anyone can be a node and interact with the network by either submitting transactions, and hence adding entries to the ledger, or participating in the process of transaction verification and block mining, or even creating smart contracts. In other words, anyone can read the chain and write a new block into the chain. In contrast, permissioned blockchains limit the parties who can transact on the blockchain and can contribute to its state. Actually, in a permissioned blockchain, only a restricted set of users have the rights to see the recorded history, to validate the block transactions, to issue transactions of their own, to participating in its consensus process, or to create smart contracts [65].

Based on the above categorization, blockchain networks can be public permissionless, public permissioned, private permissionless or private permissioned [66]. While the terms public and permissionless are used interchangeably in the relevant literature, they are actually referring to the blockchain authentication and authorization model respectively [67, 68]. Yet, for the sake of simplicity, in what follows, we also use these terms interchangeably to refer to the case of public permissionless blockchain.

Apparently, private and consortium permissioned blockchains, acting as closed ecosystems in which some central authorities control participation, cannot be regarded as fully decentralized networks since a minimum level of trust among the nodes is sustained. Instead, consortium permissioned blockchains are regarded as semi-decentralized, while private permissioned blockchains are usually compared to centralized networks, or even to distributed and centralized databases [69, 70]. Indeed, given that a permissioned blockchain shares similarities with a centralized database [70], a careful examination of the system requirements needs to be carried out before this type of blockchain is to be selected in any application scenario.

Permissionless blockchains usually employ fat protocols that compensate network contributors with tokens. On the other hand, permissioned blockchains generally do not need to employ a cryptocurrency model or monetary tokens due to the nature of these business networks. Nonetheless, both types of blockchains have their own advantages and disadvantages and can be suitable for different kind of situations. Although it may seem that in an institutional context, permissioned blockchains is unquestionably a better choice, it has been argued that permissionless blockchains operating within or across organizations still have a lot to offer [71].

5.3.1.2 Maintaining Trust Through Consensus Protocols

Even though blockchain is advertised as being trustless, trust is actually maintained in blockchains, albeit differently. As a matter of fact, while blockchain minimizes the need for certain kinds of trust, e.g. removing single points of failure, it introduces a new kind of trust, the decentralized trust that relies on the trustworthiness of the systems themselves [62]. In view of the fact that a third party is no longer needed in a blockchain to verify data integrity and to maintain trust, as opposed to the centralized architectures, consensus algorithms are used to maintain data consistency [65]. To put it another way, a consensus protocol allows all nodes of the blockchain, and the DLTs in general, to agree on a single version of the truth, i.e. on the transactions and the order in which these are listed on the newly-mined block, without the need of a trusted third party. Otherwise, the individual copies of the ledger will diverge and it will end up with branches, called forks, of the chain; the nodes will have then a different view of the global state [47]. As previously mentioned, while every node in a permissionless blockchain could take part in the consensus process, only a selected set of nodes are responsible for validating the block in a permissioned blockchain. Some of the main consensus protocols used as of today are Proof of Work (PoW), Proof of Stake (PoS), Delegated Proof-of-Stake (DPoS), Proof of Authority (PoA), and Practical Byzantine Fault Tolerance (PBFT).

In PoW, which is the underlying consensus of the bitcoin, several nodes of the distributed ledger, called miners, compete to solve a complicated mathematical problem, that is to calculate a hash value of the block header equal to or smaller than a threshold, and hence to validate a block of transactions. Once the first miner finds a solution, it broadcast it to the other nodes which then verify the solution by mutually confirming the correctness of the hash value. If all the nodes agree on the solution, the consensus is reached, and the new block is appended to all the ledgers held by the nodes of the network. The idea is that the solution to the problem is hard to find but easy to verify by the rest of the network. However, there might be cases of multiple nodes finding a solution nearly at the same time and as a result valid blocks to be generated simultaneously, resulting thereby in orphan blocks [72, 73] and in some rare cases in forks [74]. Forks typically happen to upgrade the blockchain features or to reverse the effects of hacking or catastrophic bugs and they require consensus [75]. A chain that becomes longer thereafter is judged as the authentic one, and the orphan blocks are discarded from the blockchain without compensation. Yet, in blockchains like ethereum orphan blocks are called "uncles" because they are still rewarded and added to the blockchain [76]. In fact, miners in ethereum are incentivized to include uncles in a mined block by referencing them in a new field in the header of each block. Since referencing these uncles makes the chain heavier, in ethereum, unlike bitcoin, the "heaviest" chain—and not the longest—is the authentic one. The PoW is characterized by its high energy consumption since a huge amount of computational power is required for solving the mathematical puzzle to mine a block. Moreover, in PoW, there is always the possibility of the formation of mining pools, i.e. groups of miners who pool their resources together and potentially could control the network. In PoS, which is regarded as an energy-saving alternative to PoW, miners have to prove the ownership of the amount of currency since it is believed that people with more currencies would be less likely to attack the network [65]. Although PoW is currently employed by the two biggest public permissionless blockchains, the bitcoin and the ethereum, the PoS protocol is gaining ground in public cryptocurrencies as its various nuances have been adopted by open blockchains like Cardano and DASH. In addition, ethereum is going to move to its own implementation of the PoS consensus in its highly anticipated forthcoming upgrade [77]. Nevertheless, as discussed in [78], PoS protocols may suffer several rare attacks that are highly dependent on the participation of parties having, or being able, to manage the keys of considerable stake at a given point, as well as on the lack of properly integrated countermeasures. DPoS is a more efficient PoS mechanism that uses a reputation system and real-time voting to achieve consensus. Nodes vote for representatives to secure their network and representatives are rewarded by validating transactions for the next block. In PoA, transactions are validated by approved accounts, known as validators. By attaching a reputation to an identity, validators are incentivized to uphold the transaction process, to avoid having their identities linked to a negative reputation. PBFT reach a consensus without the energy consumption required by PoW. The consensus decision is determined based on the total decisions submitted by all the nodes and the honest nodes come to an agreement of the state of the system through a majority [79].

By definition, consensus protocols in permissionless blockchains promote and establish decentralized trust in non-trusted environments. This is the result of their employed incentive mechanisms which rely mostly on game-theoretic principles for the correct operation and assume absolute non-trust among the participants. Instead, the consensus protocols operate on the assumption that all miners behave in a way that is profitable to them [80]. In an ideal scenario where there would be a minimum level of trust, all validating nodes would vote on the order of transactions for the next block, and they would go with what the majority decides [47]. However, due to the complete absence of trust in permissionless blockchains, nodes cannot rely on each other, and therefore, they are rewarded with incentives for correct behaviour to collectively agree on the state of the ledger [80]. In PoW for instance, if a malicious user tries to subvert the system by creating a fork and entering into a race with other miners to create an alternate ledger, the resulting computational cost will be tremendous, even in the case of winning. Instead, if the same work is directed towards honest mining, it can possibly result in bigger profits by way of incentives. Hence, trying to defraud the system is generally not in one's interest. This is why the trustlessness of PoW consensus mechanism, which makes no assumptions about the honesty or reliability of participants, is currently considered more suitable for permissionless blockchains. On the contrary, due to the risk of Sybil attacks public blockchains cannot rely on the PBFT consensus algorithm which requires a majority of honest nodes: even when there is only one malicious participant, it can create multiple fake identities, get multiple votes, and thus influence the network to favour its interests, forcing the number of honest nodes to a minority [47, 80].

The consensus mechanisms employed in permissioned blockchains can be the same as in permissionless networks or can be completely uniquely designed (e.g. authority-based). In fact, it has been argued that consensus-based on cryptocurrency is unsustainable for enterprise use and permissioned blockchains [81]. For instance, Hyperledger Fabric [82], a permissioned blockchain infrastructure oriented towards enterprises, does not require a built-in cryptocurrency because consensus is not reached via mining [83]. Generally speaking, given the trusted model of private permissioned blockchains and the known identities of the network participants, the consensus mechanisms used are computationally inexpensive when compared to PoW as there is no need for protection through mining. In fact, private blockchains are far less costly to operate since, as long as the majority of validators are following the rules, blocks only need a simple digital signature from the nodes that approve them instead of expensive consensus protocols [84]. Most common voting-based consensus protocols preferred by permissioned blockchains are based on the family

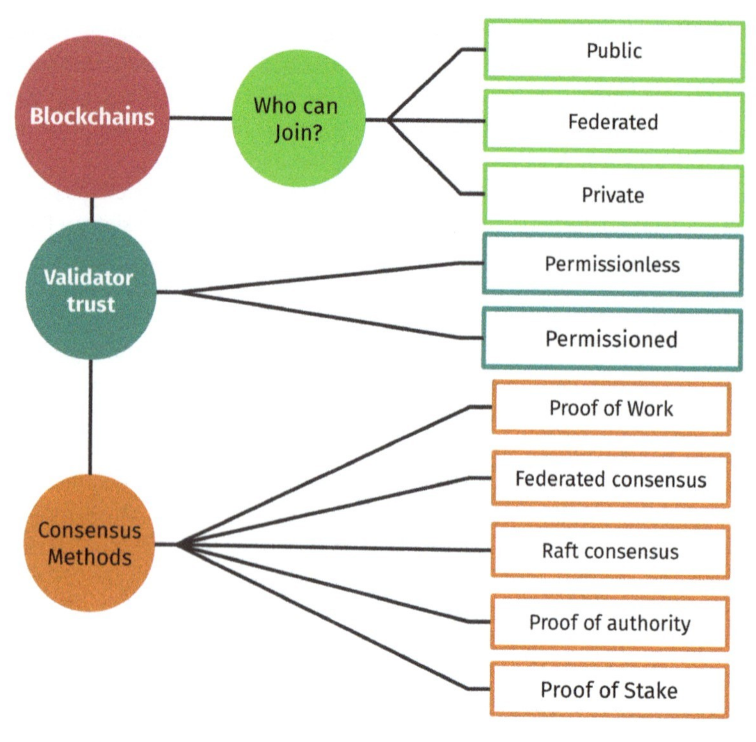

Fig. 5.1 Blockchain categorisation

protocols of Paxos and PBFT [80, 85, 86]. Given the extensive length of consensuses used in DLTs and the peculiarities of each one, only the main consensus protocols employed in blockchains were briefly mentioned here. The interested reader may further refer to [65, 85, 87]. A brief categorization of blockchains based on some of their basic characteristics is illustrated in Fig. 5.1.

5.3.1.3 Blockchain Immutability

Immutability, or irreversibility, is claimed to be a fundamental blockchain property that stems from the fact that transactions cannot be edited or deleted once they are successfully verified and recorded into the blockchain. In reality though, as we discuss below, there is not absolute immutability since there are rare cases where the records of the blockchain have been indeed reversed. However, these cases always leave evidence of tampering. Therefore, blockchains are described more accurately as tamper-evident structures instead of immutable. Tamper evidence is the consequence of the cryptographically linked blocks, which are chained together with the hash value of the preceding (parent) block. In particular, each block contains a reference to the preceding block by including in its header a cryptographic hash of the transaction data within the preceding block. This cryptographic hash is actually calculated using a Merkle tree on all the transactions of the block. A Merkle tree is a data structure constructed by recursively hashing pairs of transactions until there is only one hash, called the Merkle root. Merkle trees are used in bitcoin to summarize all the transactions in a block, producing an overall digital fingerprint of the entire set of transactions, providing thus a very efficient process to verify whether a transaction is included in a block without the need for a complete local copy of all transactions. Since the root is known and secured through the mining process, branches can be loaded on demand from untrusted sources. The cryptographic hash algorithm used in bitcoin's Merkle trees is SHA256 applied twice, also known as double-SHA256. In bitcoin blockchains, simplified payment verification (SPV) based on Merkle tree is used in order to keep the size and the computational effort low, whereas in ethereum a variation of Patricia Merkle Tree is used.

In simpler terms, the Merkle root which comprises the information from all transactions of a given block is included in the block header of the subsequent (child) block. Bearing in mind the collision-resistant property of the hash functions, any change of the transaction data in a block will change the hash of this block and, to maintain the integrity of the parent-child reference, will necessitate a change in its reference within its child block. This cascade effect ensures that once a block has many generations following it, it cannot be changed without forcing a recalculation of all subsequent blocks since such a recalculation would require enormous computation [88] for Proof of Work-based protocols. Hence, the blockchain consensus protocol guarantees the tamper-resistance nature of blockchains since tampering with already recorded data is financially detrimental. The longest the chain of blocks a blockchain has, the more resilient the blockchain is to data tampering attacks because if an adversary modifies data anywhere in the blockchain, it will result in the hash pointer in the subsequent block being incorrect, whereas any recalculation of the hashes in the subsequent blocks would be tremendously expensive [89, 90]. Therefore, for properly deployed blockchains, data residing in blockchains cannot be ever mutated or removed without leaving evidence of the tampering. Even though tampering with data already stored in the blockchain is almost impossible, data can be appended to the blockchains. Therefore, blockchains are known as append-only, tampered-proof and immutable data structures. Inevitably, since blockchain's immutability assures its transactional integrity, i.e. the correct and permanent storage of blocks and transactions within the blockchain, immutability is of paramount importance to blockchain's security and a cornerstone of its highly praised values of trustlessness and censorship-resistance.

While explained, blockchains are in principle immutable as it is nearly impossible to delete, update or rollback transactions once they are included in a blockchain, some would argue otherwise: considering that immutability is an emergent, and not intrinsic, property of a blockchain data structure, and therefore an agent or set of agents with a sufficient amount of computing power can modify it, stating that a blockchain is by default immutable is incorrect and misleading [78, 91, 92]. Especially in the context of permissioned blockchains where the number of nodes is limited, tampering with blockchain data should not be regarded as impossible since there is always a possibility of the majority of the consortium or the dominant organization nodes to vote for their version of the truth and to amend the ledger accordingly [65, 93]. Hence, although in public blockchains the existence of a long chain of blocks makes the blockchain's deep history immutable due to the extremely high cost involved for altering the hash-based integrity of the blocks, ensuring immutability in private blockchains is much cheaper and stronger, as long as the majority of validating nodes are following the rules [84, 88]. However, it has been argued that even in public blockchains there is no such thing as perfect immutability since under certain conditions a particular blockchain can be changed [84]. Although events such as the Ethereum Decentralized Autonomous Organization (DAO) fork clearly align with such claims [94], these hard forks are exceptionally rare and definitely cannot be applied on a regular basis. Hence, it is commonly held that altering transactional data in public blockchains is thus far practically impossible. For the sake of simplicity and clarity, the term "immutability" is used hereafter to denote this pragmatic unfeasibility of tampering with blockchain data on the basis of complying with regulatory and privacy requirements [91].

5.3.2 *Decentralized Storage and File Sharing*

Decentralized p2p systems for file storage and sharing, which work by sharing a file across a p2p network, they have recently been revealed as the new alternatives to traditional blockchain storage. However, as decentralized computing signs also a transition away from cloud computing, decentralized storage and file sharing is commonly discussed as an alternative to cloud storage solutions like Dropbox[6] and Google Drive[7] which rely on large, centralized silos of data. Decentralized storage and file sharing applications offer increased benefits compared to their centralized counterparts which are vulnerable to a single point of failure or to outside attacks from malicious actors who can compromise the data or leak confidential information. These events could result in adverse privacy implications, especially when the data under consideration pertain to sensitive personal details about individuals. Decentralized networks, however, do not suffer from the security limitations of centralized storage systems. For instance, due to their inherent design that relies on too many peers for securing their state, denial of service attacks to decentralized systems are less possible.

Systems such as Storj [95], SIA [96], Filecoin,[8] IPFS [97], Dat [98] and Swarm[9] are, among others, some of the most prominent decentralized p2p file sharing and storage networks [99]. Storj, SIA and Filecoin are open source platforms for decentralized storage that leverage blockchain technology and cryptocurrencies to incentivize file storage and sharing. IPFS, Dat and Swarm implement their own protocols

[6] http://www.dropbox.com.

[7] http://www.google.com/drive.

[8] http://filecoin.io/filecoin.pdf.

[9] https://swarm-guide.readthedocs.io/en/latest/introduction.html#introduction.

for efficient decentralized storage and content distribution. Yet, Swarm, which at the time of writing is still under development, is being tightly coupled with the ethereum blockchain ecosystem and its primary objective is to provide a sufficiently decentralized store for Ethereum's DApp code and data. Dat is an application protocol that focuses on sharing large files, tracking their version history automatically and supporting multiple writing.

The IPFS, on the other hand, implements a lower level more generic p2p network protocol, not tied to just one blockchain platform like ethereum. However, lacking an incentivization mechanism, as it does not support any cryptocurrency, it provides no storage guarantee. Although Filecoin is going to fill this gap by implementing an open-sourced, public cryptocurrency on top of the IPFS to incentivize users to contribute their unused storage,[10] IPFS' ambitious goal is to offer the infrastructure for reinventing the Internet and replacing the traditional HTTP protocol[11] by connecting all computing devices with the same file system. In the following paragraph, we delve into the basic characteristics of the IPFS protocol.

5.3.2.1 The IPFS Protocol

The IPFS is a p2p open-source platform, both a protocol and a network, designed to create a permanent, decentralized method of efficient and robust data storage and sharing. IPFS seeks to connect all computing devices with the same file system and to create a new way to serve information on the web [97]. Although there have been many attempts in the past to introduce decentralized file systems, IPFS is the first general file system that achieves low latency and decentralized distribution on a global scale and in the infrastructure layer. This is because IPFS provides not only the application to distribute files but also the base protocol to accomplish this.

While IPFS synthesizes successful ideas from previous p2p systems, its main contribution lies in simplifying, evolving, and connecting proven techniques into a single cohesive system, greater than the sum of its parts [97]. It combines technologies such as Distributed Hash Tables (DHT) (as implemented in the Kademlia protocol) [100] to coordinate and maintain metadata about p2p systems, and a BitTorrent inspired communication protocol, BitSwap, to coordinate networks of untrusting peers (swarms) to cooperate in distributing pieces of files to each other [101]. It also uses Self-Certified Filesystems (SFS) techniques to authenticate the server and to establish a secure communication channel to remote filesystems [102]. On top of these, IPFS builds a cryptographically authenticated data structure, similar to Git,[12] to support file versioning and efficient distribution: a Merkle Directed Acyclic Graph

[10] To this end, it employs two variations of proof-of-storage consensus mechanism, *Proof-of-Replication* (PoRep) and *Proof-of-Spacetime* (PoSt), to publicly verify that a node stores a particular file.

[11] https://www.sitepoint.com/ipfs-swarm-decentralized-content-publication-storage/.

[12] http://git-scm.com.

(DAG)[13] of immutable objects (representing files or other arbitrary data structures) with links to the cryptographic hash of the target object. The central IPFS principles of modelling all data as part of the same Merkle DAG object and addressing contents via their hashes provide useful properties such as content addressability, tamper resistance and deduplication [103]. According to these properties: all contents are always addressed by their cryptographic hashes; they are by default immutable since editing a file results to a new address (hash) of that file due to the collision-resistant property of the hash function; and duplicate files are only stored once since they always refer to the same hash. IPFS provides two protocols for creating mutable addresses to reference always the latest version of an object: IPNS and DNSLink, with the latter being more efficient [103]. Moreover, the InterPlanetary Linked Data (IPLD)[14] set of standards are implemented in the IPFS to create more flexible universally addressable and linkable decentralized data structures of different sorts of data.

In the IPFS network, nodes store a collection of objects (hashed files) in local storage, and they connect to each other to transfer objects. Nodes, i.e. users, are not required to store all the data published in the network. Instead, they can choose which data they want to persist. Users who want to retrieve any of those files access an abstraction layer where they simply call the hash of the file they want. IPFS then, after searching carefully through the nodes, takes care of finding the closest peer who has what they need and supplies the users with the file. Accessed resources are cached locally in the IPFS node to make those resources available for upload to other nodes and thereby to help with the load distribution for popular content.

As opposed to traditional location addressing used by the HTTP where a single server hosts many files and information has to be fetched by accessing this server, the content addressability, i.e. looking up the content by its cryptographic hash, ensures the authenticity of content regardless of where it is located. The implications of this property are tremendous as the IPFS could transform the Internet from being location-based, to be a content-based distributed file network. First and foremost, IPFS eliminates the HTTP problem of broken links as a given address will always point to the same content added to the IPFS network because even a slight change will result to a different address. In other words, in IPFS, the weblinks cannot break since the names of the objects are always the same. An overview of how the IPFS works is illustrated in Fig. 5.2.

As already mentioned, another powerful characteristic of IPFS is its censorship resistance since web content does not depend any more on a single entity. This censorship-resistant nature has already been exploited in many occasions to bypass web policy restrictions and to enable the freedom of speech[15] and the right to information.[16] In this regard, it has been argued that the IPFS could evolve the web and even replace the most successful "*distributed system of files*" ever deployed, the HTTP [97].

---

[13] http://docs.ipfs.io/guides/concepts/merkle-dag/.

[14] https://ipld.io/.

[15] https://www.eurekastreet.com.au/article/inside-catalonia-s-cypherpunk-referendum.

[16] https://observer.com/2017/05/turkey-wikipedia-ipfs/.

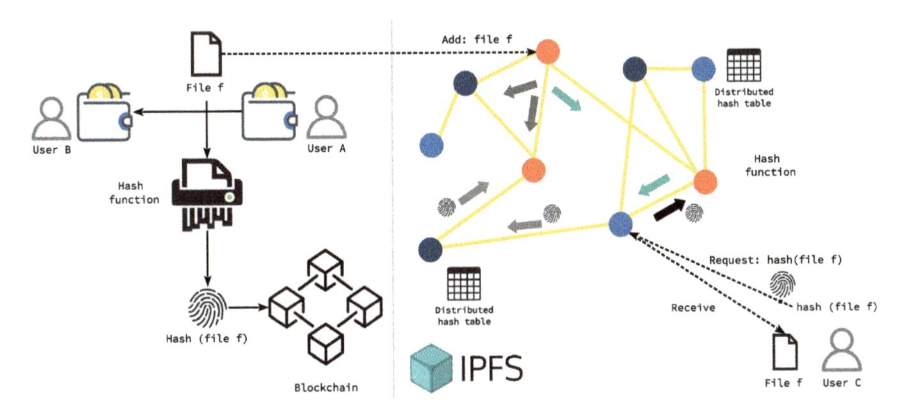

Fig. 5.2 An overview of the main IPFS operations. On the left, we depict a typical operation when data are stored in IPFS (or other DFS) and only their hashes and metadata are stored in the blockchain. On the right, a high-level overview of the IPFS data storage and retrieval is depicted. First, a user stores a file f in IPFS. Next, to retrieve these data, another user performs a request for file f using its corresponding hash value (bottom right). As the nodes are aware of the location of that file (i.e. due to the use of the DHT), they are able to efficiently retrieve the data from the nodes storing the queried file (orange nodes) to deliver it to the user finally

IPFS's substantial impact on linking and searching content online, along with its smooth integration with current blockchain platforms for storing data off-chain, contributed to its wide adoption by many blockchain projects. As a matter of fact, it is currently used by numerous projects and users who seek to share their files effortlessly and efficiently. On top, big corporations, like Cloudflare, leveraged IPFS to provide their cloud services.[17]

References

1. *IBM's Plans to Ship Simon Put on Hold for Time Being* (1994). http://research.microsoft.com/en-us/um/people/bibuxton/buxtoncollection/a/pdf/press%20release%20delay%201994.pdf. Accessed 15-02-2017
2. N. Lathia, V. Pejovic, K.K. Rachuri, C. Mascolo, M. Musolesi, P.J. Rentfrow, Smartphones for large-scale behavior change interventions. IEEE Pervasive Comput. **12**(3), 66–73 (2013)
3. N.D. Lane, E. Miluzzo, H. Lu, D. Peebles, T. Choudhury, A.T. Campbell, A survey of mobile phone sensing. IEEE Commun. Mag. **48**(9), 140–150 (2010)
4. E. Miluzzo, Smartphone sensing. Ph.D. thesis, Dartmouth College Hanover, New Hampshire (2011)
5. E. Macias, A. Suarez, J. Lloret, Mobile sensing systems. Sensors **13**(12), 17292–17321 (2013)
6. M. Weiser, Ubiquitous computing. Computer **10**, 71–72 (1993)
7. M. Weiser, The computer for the 21st century. Sci. Am. **265**(3), 94–104 (1991)
8. M. Raento, A. Oulasvirta, N. Eagle, Smartphones an emerging tool for social scientists. Sociol. Methods Res. **37**(3), 426–454 (2009)

[17] https://blog.cloudflare.com/distributed-web-gateway/.

9. D. Zhang, B. Guo, Z. Yu, The emergence of social and community intelligence. Computer **44**(7), 21–28 (2011)
10. G. Chittaranjan, J. Blom, D. Gatica-Perez, Mining large-scale smartphone data for personality studies. Pers. Ubiquit. Comput. **17**(3), 433–450 (2013)
11. N. Eagle, A.S. Pentland, D. Lazer, Inferring friendship network structure by using mobile phone data. Proc. Natl. Acad. Sci. **106**(36), 15274–15278 (2009)
12. J. Hernandez, D.J. McDuff, R.W. Picard, Biophone: physiology monitoring from peripheral smartphone motions, in *2015 37th Annual International Conference of the IEEE Engineering in Medicine and Biology Society (EMBC)*, (IEEE, 2015), pp. 7180–7183
13. E. Horvitz, D. Mulligan, Data, privacy, and the greater good. Science **349**(6245), 253–255 (2015)
14. E. Politou, E. Alepis, C. Patsakis, A survey on mobile affective computing. Comput. Sci. Rev. **25**, 79–100 (2017)
15. R.W. Picard, R. Picard, *Affective Computing*, vol. 252 (MIT Press Cambridge, 1997)
16. R.W. Picard, Toward computers that recognize and respond to user emotion. IBM Syst. J. **39**(3.4), 705–719 (2000)
17. M. Rouse, *Affective Computing* (2005). http://whatis.techtarget.com/definition/affective-computing
18. J. Tao, T. Tan, Affective computing: a review, in *International Conference on Affective Computing and Intelligent Interaction* (Springer, 2005), pp. 981–995
19. R.W. Picard, S. Papert, W. Bender, B. Blumberg, C. Breazeal, D. Cavallo, T. Machover, M. Resnick, D. Roy, C. Strohecker, Affective learning–a manifesto. BT Technol. J. **22**(4), 253–269 (2004)
20. S.D. Gosling, W. Mason, Internet research in psychology. Annu. Rev. Psychol. **66**, 877–902 (2015)
21. G. Chittaranjan, J. Blom, D. Gatica-Perez, Who's who with big-five: analyzing and classifying personality traits with smartphones, in *2011 15th Annual International Symposium on Wearable Computers* (IEEE, 2011), pp. 29–36
22. R. de Oliveira, A. Karatzoglou, P. Concejero Cerezo, A. Armenta Lopez de Vicuña, N. Oliver, Towards a psychographic user model from mobile phone usage, in *CHI'11 Extended Abstracts on Human Factors in Computing Systems* (ACM, 2011) pp. 2191–2196
23. Y.A. de Montjoye, J. Quoidbach, F. Robic, A.S. Pentland, Predicting personality using novel mobile phone-based metrics, in *International Conference on Social Computing, Behavioral-Cultural Modeling, and Prediction* (Springer, 2013), pp. 48–55
24. K.K. Rachuri, M. Musolesi, C. Mascolo, P.J. Rentfrow, C. Longworth, A. Aucinas, Emotion-Sense: a mobile phones based adaptive platform for experimental social psychology research. in *Proceedings of the 12th ACM International Conference on Ubiquitous Computing* (ACM, 2010), pp. 281–290
25. H. Lee, Y.S. Choi, S. Lee, I. Park, Towards unobtrusive emotion recognition for affective social communication, in *2012 IEEE Consumer Communications and Networking Conference (CCNC)*, (IEEE, 2012), pp. 260–264
26. H.J. Kim, Y.S. Choi, Exploring emotional preference for smartphone applications, in *2012 IEEE Consumer Communications and Networking Conference (CCNC)*, pp 245–249 (IEEE, 2012)
27. A. Bogomolov, B. Lepri, F. Pianesi, Happiness recognition from mobile phone data, in *2013 International Conference on Social Computing* (IEEE, 2013), pp 790–795
28. M. Pielot, T. Dingler, J.S. Pedro, N. Oliver, When attention is not scarce-detecting boredom from mobile phone usage, in *Proceedings of the 2015 ACM International Joint Conference on Pervasive and Ubiquitous Computing* (ACM, 2015), pp. 825–836
29. R. LiKamWa, Y. Liu, N.D. Lane, L. Zhong, Can your smartphone infer your mood, in *Phone-Sense Workshop* (2011), pp 1–5
30. E. Ceja, V. Osmani, O. Mayora, *Automatic Stress Detection in Working Environments from Smartphones' Accelerometer Data: A First Step* (IEEE J. Biomed, Health Inform, 2015)

31. H. Lu, D. Frauendorfer, M. Rabbi, M.S. Mast, G.T. Chittaranjan, A.T. Campbell, D. Gatica-Perez, T. Choudhury, StressSense: detecting stress in unconstrained acoustic environments using smartphones, in *Proceedings of the 2012 ACM Conference on Ubiquitous Computing* (ACM, 2012), pp. 351–360
32. G.C.L. Hung, P.C. Yang, C.C. Chang, J.H. Chiang, Y.Y. Chen, Predicting negative emotions based on mobile phone usage patterns: an exploratory study. JMIR Res. Protocols **5**(3) (2016)
33. P. Bardhan, Decentralization of governance and development. J. Econ. Perspect. **16**(4), 185–205 (2002)
34. P. Seabright, Accountability and decentralisation in government: an incomplete contracts model. Eur. Econ. Rev. **40**(1), 61–89 (1996)
35. P. Baran, On distributed communications networks. IEEE Trans. Commun. Syst. **12**(1), 1–9 (1964)
36. *Social Media Alternatives Series, ep. 1: What You Need to Know* (2018). https://www.fliphodl.com/social-media-alternatives-series-ep-1-what-you-need-to-know/
37. D.S. Milojicic, et al., Peer-to-peer computing. Technical Report HPL-2002-57, HP Labs (2002)
38. Cm.A. Yeung, I. Liccardi, K. Lu, O. Seneviratne, T. Berners-Lee, Decentralization: the future of online social networking, in *W3C Workshop on the Future of Social Networking Position Papers*, vol 2 (2009), pp. 2–7
39. S. Nakamoto, et al., Bitcoin: a peer-to-peer electronic cash system (2008)
40. A. Deshpande, K. Stewart, L. Lepetit, S. Gunashekar, Distributed ledger technologies/blockchain: challenges, opportunities and the prospects for standards. Overview report The British Standards Institution (BSI) (2017)
41. N. El Ioini, C. Pahl, A review of distributed ledger technologies, in *OTM Confederated International Conferences" on the Move to Meaningful Internet Systems"* (Springer, 2018), pp 277–288
42. S. Haber, W.S. Stornetta, How to time-stamp a digital document, in *Conference on the Theory and Application of Cryptography* (Springer, 1990), pp. 437–455
43. F. Casino, T.K. Dasaklis, C. Patsakis, A systematic literature review of blockchain-based applications: current status, classification and open issues. Telemat. Inform. (2018)
44. P. Basson, *The Untold History of Bitcoin: Enter the Cypherpunks* (2018). https://medium.com/swlh/the-untold-history-of-bitcoin-enter-the-cypherpunks-f764dee962a1
45. S. Falkon, *Cypherpunks and the Rise of Cryptocurrencies* (2017a). https://medium.com/swlh/cypherpunks-and-the-rise-of-cryptocurrencies-899011538907
46. L. Law, S. Sabett, J. Solinas, *How to Make a Mint: The Cryptography of Anonymous Electronic Cash* (1996). http://groups.csail.mit.edu/mac/classes/6.805/articles/money/nsamint/nsamint.htm
47. K. Christidis, M. Devetsikiotis, Blockchains and smart contracts for the internet of things. IEEE Access **4**, 2292–2303 (2016)
48. G. Greenspan, *Ending the Bitcoin vs Blockchain Debate* (2015). https://www.multichain.com/blog/2015/07/bitcoin-vs-blockchain-debate/
49. *CoinMarketCap, Cryptocurrency Market Capitalizations* (2019). https://coinmarketcap.com/
50. M. Swan, *Blockchain: Blueprint for a New Economy* (O'Reilly Media, Inc., 2015)
51. V. Buterin, Ethereum white paper: a next generation smart contract & decentralized application platform. First version (2014)
52. Bitcoin Exchange Guide News Team, *New Study Shows 43 Blockchain Solutions Implemented had Zero Percent Success Rate* (2018). https://bitcoinexchangeguide.com/new-usaid-study-shows-43-blockchain-solutions-implemented-had-no-success-rate/
53. D. Gerard, *Blockchain Identity: Cambridge Analytica, But on the Blockchain* (2018). https://davidgerard.co.uk/blockchain/2018/03/22/blockchain-identity-cambridge-analytica-but-on-the-blockchain/
54. K. Stinchcombe, *Blockchain is Not Only Crappy Technology But a Bad Vision for the Future* (2018). https://medium.com/@kaistinchcombe/decentralized-and-trustless-crypto-paradise-is-actually-a-medieval-hellhole-c1ca122efdec

55. T.K. Dasaklis, F. Casino, C. Patsakis, Blockchain meets smart health: towards next generation healthcare services, in *2018 9th International Conference on Information. (Intelligence, Systems and Applications (IISA)* (2018), pp. 1–8
56. K.R. Lakhani, M. Iansiti, *The Truth About Blockchain*. Harvard Business Review Harvard University Retrieved, pp. 01–17 (2017)
57. L. Mearian, *What is Blockchain? The Complete Guide* (2019). https://www.computerworld.com/article/3191077/what-is-blockchain-the-complete-guide.html
58. Z. Zheng, S. Xie, H.N. Dai, H. Wang, Blockchain challenges and opportunities: a survey. Work Pap-2016 (2016)
59. *Bitcoin as Digital Gold* (2018). https://www.luno.com/learn/en/article/bitcoin-as-digital-gold
60. J.M. Duffy, *Ethereum Will be the Backbone of the New Internet* (2018). https://medium.com/loom-network/ethereum-will-be-the-backbone-of-the-new-internet-88718e08124f
61. S. Marsh, M.R. Dibben, Trust, untrust, distrust and mistrust–an exploration of the dark (er) side, in *International Conference on Trust Management* (Springer, 2005). pp. 17–33
62. K. Werbach, *The Blockchain and the New Architecture of Trust* (MIT Press, 2018)
63. A. Panarello, N. Tapas, G. Merlino, F. Longo, A. Puliafito, Blockchain and Iot integration: a systematic survey. Sensors **18**(8), 2575 (2018)
64. Y. Vilner, *No More Hype: Time To Separate Crypto From Blockchain Technology* (2018). https://www.forbes.com/sites/yoavvilner/2018/11/14/no-more-hype-time-to-separate-crypto-from-blockchain-technology/#1d5ea73b171c
65. S. Meiklejohn, Top ten obstacles along distributed ledgers path to adoption. IEEE Secur. Privacy **16**(4), 13–19 (2018)
66. G. Hileman, M. Rauchs, 2017 global blockchain benchmarking study (2017)
67. *8202 NIR, Blockchain Technology Overview* (2018). https://doi.org/10.6028/NIST.IR.8202
68. R. Houben, A. Snyers, *Cryptocurrencies and Blockchain: Legal Context and Implications for Financial Crime, Money Laundering and Tax Evasion* (European Parliament, Bruxelles, 2018)
69. *How are Permissioned/Private Blockchains Immutable or Better Than a Database*? (2018). https://www.reddit.com/r/BitcoinBeginners/comments/8zw7ce/how_are_permissionedprivate_blockchains_immutable/
70. K. Wüst, A. Gervais, Do you need a blockchain? in *2018 Crypto Valley Conference on Blockchain Technology (CVCBT)* (IEEE, 2018), pp. 45–54
71. V. Buterin, *On Public and Private Blockchains* (2015). https://blog.ethereum.org/2015/08/07/on-public-and-private-blockchains/
72. T. Cryptonomist, *An Orphan Block on the Bitcoin (BTC) Blockchain* (2019). https://en.cryptonomist.ch/2019/05/28/orphan-block-bitcoin-btc-blockchain/
73. B.B. Explorer, *Orphaned Blocks* (2019). https://bitcoinchain.com/block_explorer/orphaned
74. *Bitcoin Fork and Bitcoin Hard Fork List* (2019). https://www.forks.net/list/Bitcoin/
75. S. Falkon, *The Story of the DAO—Its History and Consequences* (2017b). https://medium.com/swlh/the-story-of-the-dao-its-history-and-consequences-71e6a8a551ee
76. Etherscan, *Ethereum Uncles Count and Rewards Chart* (2019). https://etherscan.io/chart/uncles
77. *Ethereum 2.0: Proof of Stake (PoS)* (2019). https://docs.ethhub.io/ethereum-roadmap/ethereum-2.0/proof-of-stake/
78. E. Deirmentzoglou, G. Papakyriakopoulos, C. Patsakis, A survey on long-range attacks for proof of stake protocols. IEEE Access **7**, 28712–28725 (2019). https://doi.org/10.1109/ACCESS.2019.2901858
79. G.T. Nguyen, K. Kim, A survey about consensus algorithms used in blockchain. J. Inf. Process. Syst. **14**(1) (2018)
80. R. Thurimella, Y. Aahlad, *The Hitchhiker's Guide to Blockchains: A Trust Based Taxonomy* (2018). https://wandisco.com/assets/whitepapers/the-hitchhikers-guide-to-blockchains.pdf
81. T. Jenks, *Pros and Cons of Different Blockchain Consensus Protocols* (2018a). https://www.verypossible.com/blog/pros-and-cons-of-different-blockchain-consensus-protocols

82. *Hyperledger Fabric* (2021). https://www.hyperledger.org/projects/fabric
83. T. Jenks, *Using Blockchain Technology in Your Project: The Ultimate Guide to Building a Blockchain Application* (2018b). https://www.verypossible.com/using-blockchain-technology-in-your-project
84. G. Greenspan, *The Blockchain Immutability Myth* (2017). https://www.multichain.com/blog/2017/05/blockchain-immutability-myth/
85. C. Cachin, M. Vukolić, Blockchain consensus protocols in the wild (2017). arXiv preprint arXiv:170701873
86. L. Lamport, Paxos made simple. ACM Sigact News **32**(4), 18–25 (2001)
87. L.S. Sankar, M. Sindhu, M. Sethumadhavan, Survey of consensus protocols on blockchain applications, in *2017 4th International Conference on Advanced Computing and Communication Systems (ICACCS)* (IEEE, 2017), pp. 1–5
88. A.M. Antonopoulos, *Mastering Bitcoin: Unlocking Digital Cryptocurrencies* (O'Reilly Media, Inc. 2014)
89. A. Narayanan, J. Bonneau, E. Felten, A. Miller, S. Goldfeder, *Bitcoin and Cryptocurrency Technologies: A Comprehensive Introduction* (Princeton University Press, 2016)
90. F. Tschorsch, B. Scheuermann, Bitcoin and beyond: a technical survey on decentralized digital currencies. IEEE Commun. Surv. Tutor. **18**(3), 2084–2123 (2016)
91. M. Finck, Blockchains and data protection in the European Enion. Eur Data Prot. L Rev. **4**, 17 (2018)
92. D. Conte de Leon, A.Q. Stalick, A.A. Jillepalli, M.A. Haney, F.T. Sheldon, Blockchain: properties and misconceptions. Asia Pac. J. Innov. Entrepreneurship **11**(3), 286–300 (2017)
93. T. Swanson, Consensus-as-a-service: a brief report on the emergence of permissioned, distributed ledger systems (2015)
94. D. Siegel, *Understanding The DAO Hack for Journalists* (2016). https://medium.com/@pullnews/understanding-the-dao-hack-for-journalists-2312dd43e993
95. S. Wilkinson, T. Boshevski, J. Brandoff, V. Buterin, Storj a peer-to-peer cloud storage network (2014)
96. D. Vorick, L. Champine, Sia: *Simple Decentralized Storage* (2014). White paper available at https://siatech/siapdf
97. J. Benet, IPFS-content addressed, versioned, p2p file system (2014). arXiv preprint arXiv:14073561
98. M. Ogden, Dat-distributed dataset synchronization and versioning. OSF Preprints (2017)
99. F. Casino, E. Politou, E. Alepis, C. Patsakis, Immutability and decentralized storage: an analysis of emerging threats. IEEE Access **8**, 4737–4744 (2020)
100. P. Maymounkov, D. Mazieres, Kademlia: A peer-to-peer information system based on the xor metric, in *International Workshop on Peer-to-Peer Systems* (Springer, 2002), pp. 53–65
101. B. Cohen, Incentives build robustness in bittorrent. Workshop Econ. Peer-to-Peer Syst. **6**, 68–72 (2003)
102. D.F. Mazières, Self-certifying file system. Ph.D. thesis, Massachusetts Institute of Technology (2000)
103. C. Patsakis, F. Casino, *Hydras and Ipfs: A Decentralised Playground for Malware* (Int. J. Inf, Secur, 2019)

Chapter 6
Privacy in Ubiquitous Mobile Computing

Abstract Mobile sensing applications exploit big data to measure and assess human-behavioural modelling. However, big data profiling and automated decision practices, albeit powerful and pioneering, they are also highly unregulated and thereby unfair and intrusive. Their risk to privacy has been indeed identified as one of the biggest challenges faced by mobile computing applications. In this Chapter, we delve into the privacy risks arising from the ubiquitous mobile computing and sensing applications and, in particular, from the big data algorithmic processing which infers sensitive personal details such as people's social behaviour or emotions. In this respect, we thoroughly discuss the risks of profiling which are further elaborated in the tax and financial context. We also explore strategies towards mitigating these privacy risks, and we investigate the extent to which the GDPR protects against these threats, especially against aggressive profiling and automated decision-making. To mitigate these risks, we explore implementation challenges and we introduce countermeasures in the context of financial privacy that can adhere to the privacy requirements of the GDPR.

6.1 Introduction

From the early days of ubiquitous and affective computing in the 90s up to today where mobile sensing applications exploit big data to measure and assess human-behavioural modelling, the progress in the field of mobile computing and HCI is remarkable [6, 24]. Artificial intelligence (AI), ubiquitous mobile sensing, and HCI are rapidly evolving and many advocate that emotional-aware smartphones will be a commonplace in the days to come. Despite these radical developments, mobile affect recognition systems face a number of substantial challenges, both ethical and technical. And although technology constantly evolves, providing evidence that technical challenges will be confuted at some point, the discussion on their ethical challenges are still in their infancy [34, 110]. Miller in its smartphone psychology manifesto [95] highlights that, *"one of the main disadvantages of smartphones are the ethical challenges in obtaining truly informed consent, protecting participant privacy and anonymity and reducing liability risks"*, while O'Hara stresses [100] that "*precau-*

E. Politou et al., *Privacy and Data Protection Challenges in the Distributed Era*, Learning and Analytics in Intelligent Systems 26,
https://doi.org/10.1007/978-3-030-85443-0_6

tions against [smartphones'] misuse were not discussed widely". Therefore, setting a privacy baseline and providing a common framework for performing all the associated research is of utter importance to our well-being.

In the same spirit, Ohm argues [102] that the four recent trends of our high tech society, namely smartphone, cloud, social networks and big data, taken together, enable the rise of a powerful, new surveillance society which raises significant new threats to privacy, such as location tracking. Although the concept of surveillance society and its impact on privacy has appeared in many non-fiction books even before the big data and smartphone era [20, 128], there are the last technological advancements and the prevalence of pervasive and ubiquitous computing, which continuously collects huge amount of personal data from online activities and mobile devices, that triggered massive skepticism and worldwide dispute around the notion of privacy [135]. Against this background, privacy has been indeed identified as one of the biggest challenges faced by mobile computing applications, either in affective or any other domain.

In this Chapter we present and discuss the risks to privacy imposed by ubiquitous mobile computing practice and research when combined with big data algorithmic processing, and we discuss their implications for individuals on the basis of the GDPR and the RtbF. We specifically delve into the risks of profiling which are further elaborated in the tax and financial context. To this respect, we consider the recent international and European policies towards financial and tax transparency, along with their implications for privacy when used for profiling and automated decision purposes. Furthermore, in paragraph 6.2.6, we explore the academic discussions towards the accountability and transparency of profiling and automated decision processing. Afterwards, we identify strategies towards mitigating these risks, investigating also the extent to which the GDPR provisions establish a protection regime for individuals against the aforementioned privacy risks and the advanced profiling practices and discriminatory automated decisions. Finally, we discuss the implementation challenges of current privacy enhancing technologies and we propose some countermeasures in the context of financial privacy.

6.2 Privacy Risks in Mobile Computing

Solove in his taxonomy of privacy [130] explains that *"all methods and practices adopted nowadays by personal data markets, such as aggregation of personal data, increased accessibility, re-identification, secondary use, exclusion, and decisional interference, constitute privacy breaches"*. However, according to Acquisti [1], the market for personal data and the market for privacy, are actually two sides of the same coin who contribute considerably to most privacy issues originated today. Admittedly, even though privacy in mobile devices is a rather complicated issue [133], respecting and preserving user's privacy is perhaps the most essential task for all mobile data mining systems.

People are usually sensitive when their personal data are captured and used, especially if the data reveal their location, communications or other sensitive information. This concern grows dramatically when health or affective data are collected or are being inferred, even when these data are to be used for research purposes only. In a pilot study led in 2016 in Australia regarding preliminary insight into individuals' perceptions towards sharing their personal health data with researchers [25], the majority of the participants required privacy assurance in order to donate their personal data to a public scientific database. Moreover, anonymity of personal health data was regarded as very or extremely important by nearly 90% of participants and it was the primary constraint on the sharing of health data.

This reluctance is not surprising, given the fact that many studies on mobile privacy have revealed serious threats in disclosing personal information. For example in [47] it was shown that the majority of applications leaked the device ID, an information which can provide detailed information about the habits of a user. Plus, there is always the possibility that additional data are used to tie a device ID to a person, increasing the privacy risks. Most certainly, releasing a smartphone application to the public requires compliance with ethical codes as well as privacy and security constraints that protect users. However, as we have shown in [105], the majority of the analysed m-health applications do not follow well-known security and privacy practices and guidelines, not even the legal restrictions imposed by the GDPR, thus jeopardizing the privacy of millions of users. Likewise, the majority of the referenced research works presented in our survey [111] do not consider privacy and security constraints at all in their design, and even those that do take some steps to meet these requirements, their undertaken measures are not as satisfactory as it would have been expected.

Undeniably, data derived from mobile sensors can reveal information that most users wouldn't have exposed willingly if they had been asked, like call and SMS conversations, photographs and location data. This continuous and passive collection of personal information from smartphone devices could be considered by many privacy advocates as a clear personal intrusion and a major threat to human rights. But not all kinds of mobile data contribute equally to the potential threats. In Fig. 6.1 all the smartphone data sources, utilized by the referenced works in [111] for inferring affective states, are taxonomized based on the risk they entail in disclosing sensitive personal information and are classified according to their privacy invasiveness into low, medium or high risks. Having said that, it should be stated that not all the studies referenced in [111] handle smartphone data sources with the same degree of invasiveness. For instance, some applications use call logs to calculate aggregated information, such as the number of calls and their duration, while others access more sensitive information such as the contacts list to calculate the diversity of social communication by counting the time allocated to each contact.

Beyond the aforementioned privacy concerns, issues of privacy are raised not only regarding the smartphones' holders but regarding others in their vicinity due to some sensors' high sensitivity, like camera and microphone [54]. In many cases, people not carrying on an mobile sensing application in their smartphones, may be monitored

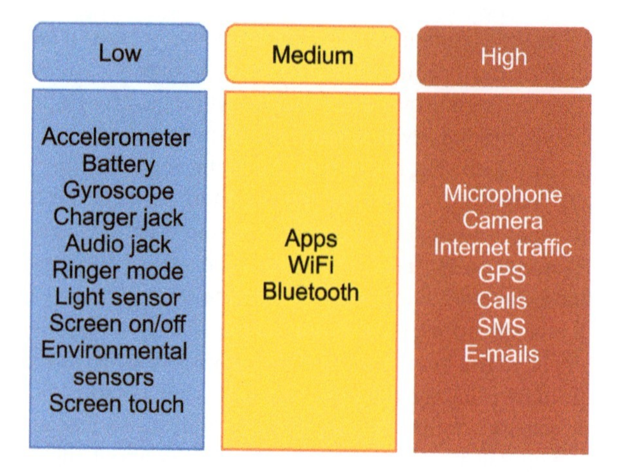

Fig. 6.1 Taxonomy of sensors and utilities according to privacy invasiveness

by just sitting next to others who they do carry on a mobile sensing application or by calling/messaging them, and thus their personal data are acquired as well.

Another interesting aspect of privacy in smartphone sensing applications is related to those data that do not only describe the current state of the individuals, but also their predicted future states [108, 109]. This raises a question pointed out by many researchers regarding the ownership of the information extracted and inferred from the processing of one's personal data. Besides, as we elaborate in Sect. 6.2.4, the impact of the inferred personal information when this is used for profiling and automated decision-making can be devastating.

Below, we delve into the most serious privacy risks arising from the expansion of mobile computing and the big data algorithmic processing.

6.2.1 *Privacy and Big Data*

The emerging areas of pervasive applications characterize the boom of big data era, where sheer volume of data are collected and processed to promote machine's intelligence by learning via example. For instance, as explained in Chap. 5, the continuous capture of information coming from sensor measurements, like accelerometer, Bluetooth, gyroscope etc., along with the way that an individual uses its smartphone, like number/duration/diversity of calls/SMS, ringtone used etc., when combined with intelligent machine learning (ML) methods, can infer holders' affective statuses, psychological conditions or other personal information [69]. As fascinating as this research area may sound in terms of its technological and scientific achievements, it conceals severe implications and risks to human privacy and data protection rights as personal and sensitive inferences are extracted from allegedly "innocent" data [104]. Inevitably, privacy and ethical issues arising from the use of current pervasive applications and big data exploitation have been discussed in many scientific papers

[14, 69, 96]. As we elaborate below in 6.2.4, the vast amount of personal data that are either publicly available in social networks or can be easily extracted through the continuous personal traces left behind when one is surfing the web or using ubiquitous devices, such as smartphones, can be collected and exploited for profiling and marketing purposes or even experimental research. Except the monetary exploitation, the processing of this sheer volume of personal data may raise concerns about the conditions under which they were collected, processed and disseminated.

Apparently, privacy and big data are in many cases contradictory. Big data require massive amount of information to be collected with not a predefined and clear purpose at the time of collection. Users do not have any control on their personal information stored and analysed by the involved data controllers and the parties that participate in data dissemination may be numerous [69, 96, 151]. Although there are existing approaches which can help with the issue of privacy (e.g., cryptography, privacy-preserving data mining, anonymization), they are often insufficient both in terms of technological and in operational issues [75]. For instance, collecting fine-grained personal data like those captured almost continuously by smartphone sensors can potentially lead to further privacy exposure since many things that might seem irrelevant now, when correlated with other data and context, may reveal a lot of sensitive information.

6.2.2 Informed Consent

Informed consent is a concept highly interrelated to privacy and data protection principles as it is required in cases where massive amount of personal, and especially sensitive data, is to be collected from the participants. And even then, the reuse of the collected data needs to be limited to preserve the previously acquired informed consent [114]. However, as we analysed in 3.4, informed consent and big data analysis are in many aspects contradictory to each other since the value of the captured big personal data usually reside in future and unanticipated uses. Therefore, upfront notice is not possible when the value of personal information is not apparent at the time of collection when consent is normally given.

For instance, in applications developed for research purposes, subjects are not always aware of the goals of the study and may not fully comprehend the actual data collected. In most of the cases, informed consent is something that researchers intentionally skip during their experiments in order to improve their quality and to avoid a bias of people attitudes and interaction styles. People are expected to act differently when they know that they are being monitored or surveilled [12, 56] and they do not reveal their actual emotions as they might feel that they could be judged about their reactions. As discussed in 3.4.2, the GPDR allows the notion of a broad consent in research studies whereas, in some few cases, GDPR permits researchers not to obtain consent at all. Nevertheless, when research concern sensitive information such as affective or health data, there should be a flexible balance between specific consent, which strangulates research, not consent, which trespasses individual's privacy or another type of consent.

6.2.3 *Risk of Re-Identification*

Sweeney, the pioneer of de-anonymization techniques, proved many years ago that 87% of the population in the USA had reported characteristics that likely made them unique based only on ZIP, gender and date of birth [136]. More recently, researchers uncovered the identities of sample donors using free, publicly accessible Internet resources like recreational genetic genealogy databases [58]. Although a malicious adversary can use personally identifiable data—such as those used in the aforementioned research works—to link data to identity, the adversary can do the same using information that nobody would classify as personally identifiable. For example, researchers demonstrated that the re-identification of individuals is possible based on anonymous film ratings of 500 000 subscribers of Netflix [98].

These developments led to the "*failure of anonymization*" [101], which along with big tech giants' data aggressive policies, eliminated long time ago any confidence that personal data are safe and protected. For example, in 2016 Google, the biggest online search engine, quietly changed its privacy policy to allow individuals browsing habits to be combined with what the company learns from the use of gmail and other tools,[1] a fact that has been criticized by many Consumer Protection boards as the genesis of the "super-profiles" that enable advertisers to relate personally identifiable user information with an individual's online history.[2] This phenomenon has been further amplified by the consolidation of off-line credit card transactions as well.[3]

In this perspective, the spread of ubiquitous and pervasive computing—which continuously collects huge amount of personal data from online activities and mobile devices—as well as of big data analytics, intensify additionally the threat of re-identification. As Narayanan and Shmatikov state in [99], "*big data guarantees the key precondition for achieving re-identification, namely the corresponding data attributes to be sufficiently numerous and fine-grained in way that no two people are similar, except with a small probability*". As a result, re-identification by using and combining various available sources of information, not necessary personal ones, has been characterized as one of the major privacy threats in our modern data driven society. For instance, it has been shown in many works that there is no such thing like a foolproof anonymization [93, 99, 101] since almost all information can be defined

[1] http://njtoday.net/2016/11/17/google-quietly-dropped-ban-personally-identifiable-web-tracking/.

[2] http://www.csmonitor.com/Technology/2016/1220/Privacy-groups-Serial-offender-Google-deceived-consumers-with-2016-policy-change.

[3] http://www.zdnet.com/article/google-well-track-your-offline-credit-card-use-to-show-that-online-ads-work/.

as "personal" when combined with enough other relevant data [138]. In particular, as it is explained in [99], any information that distinguishes one person from another, like consumption preferences or call usage patterns, can be used for re-identifying anonymous data (thus de-anonymizing them).

6.2.4 *Risk of Profiling*

Apart from the risk of re-identification, big data exploitation contributes significantly to the privacy risks associated with the aggressive profiling practices. Below we examine the academic literature on profiling and the impact of big data algorithmic processing in building predictive personalized profiles.

6.2.4.1 What Is Profiling

Profiling, according to its many definitions in various business and grammar dictionaries, is "*the recording and analysis of a person's psychological and behavioural characteristics, so as to assess or predict their capabilities in a certain sphere or to assist in identifying a particular subgroup of people*"[4] or "*the act or process of extrapolating information about a person based on known traits or tendencies*".[5] In other words, profiling is all about *personalization*, which according to Cohen [30] is the new religion of the information society whose high priests are the "*quant jocks*" of big data.

The academic literature on the definition of profiling is prolific with diverse interpretations under various technical, social and legal contexts, mainly due to the fact that profiling is a highly evocative term with multiple meanings, used in both specialist and non-specialist contexts [17]. Hildebrandt defines profiling as a process of "discovering" correlations between data in databases that can be used to identify and represent a human or non-human subject (individual or group) and/or the application of profiles (sets of correlated data) to individuate and represent a subject or to identify a subject as a member of a group or category [63]. In other words, profiling is a technique to automatically process personal and non-personal data, aimed at developing predictive knowledge from data in the form of constructing profiles, that is discovering unexpected patterns and probabilities between data in large data sets that can subsequently be applied as a basis for decision-making [122]. In technical terms, profiling can be understood as a specific data mining method. In this perspective, profiling is regarded as an automated process to examine large data sets in order to build classes or categories of characteristics. These can be used to generate profiles of individuals, groups, places, events or whatever is of interest aiming at generating prognostic information to anticipate future trends and to forecast behaviour,

---

[4] https://www.igi-global.com/dictionary/profiling/23752.

[5] https://www.merriam-webster.com/dictionary/profiling.

processes or developments [17] as well as to assess the risks and/or opportunities of individual subjects [63].

Several distinctions of profiling have emerged such as the distinction between group and individual, or personalized, profiling. The first categorizes and classifies groups of people based on specific characteristics, such as the affective classifications described in our survey in which state-of-the-art mobile computation methods of classifying people to specific emotional states or personality traits are analysed [111], whereas the second mines the data of one individuated subject, such as the case of behavioural biometrics [63]. Furthermore, Hildebrandt defines distributive profiling (as opposed to non-distributive) as the case where a group of which all members share all the attributes of the group's profile and hence the group profile can be applied without any problem to a member of the group, building thus a kind of personal profile [63]. Likewise, she describes automated profiling, where profiles are generated and applied in the process of data mining after which human experts filter the results before making decisions, versus autonomic machine profiling, where the decisions routinely follow the machine's "advice" without requiring a human intervention [63].

6.2.4.2 Profiling and Big Data Analytics

Undeniably, the emergence of big data contributed greatly in the data mining techniques, especially those aiming towards profiling. As reported by the UK ICO [145], big data analytics, characterized mainly by the use of ML algorithms and the huge collections of either new types of data or often repurposed, can bring benefits to businesses, to society and to individuals as consumers and citizens. Big data analytics can also help the public sector to deliver more effective and efficient services and to produce positive outcomes that improve the quality of people's lives [145]. At the same time, however, big data analytics create a new digital landscape since the predictive nature of the extracted inferences as well as the complexity and the obscurity of data processing distinguish them from previous profiling solutions [89].

Indeed, big data analytics, by combining algorithms and information from large and diverse datasets, create a new kind of knowledge as they locate unexpected and previously unknown structures, correlations and patterns [65, 106]. Thereby, a person's online and offline activities are turned into profiling scores whereas predictive algorithms mine personal information to make guesses about individuals' likely actions and behaviours. Beyond any doubt, big data analytics along with ML algorithms enable, now more than ever, the extensive profiling, namely the construction of detailed personal profile of one's life ready to be used and exploited either by private firms for profit, most commonly through advertisements or domain specific scoring systems, or by public authorities for accomplishing their duties regarding conformance audits and controls.

The risks of profiling opportunities arising due to the big data techniques are being discussed increasingly throughout the academic literature. Although Hildebrandt identified almost a decade ago the threats of profiling as commonly related to the key aspects of fundamental citizen rights, such as the rights to privacy, data protection and non-discrimination, democracy, autonomy, and self-determination, big data and algorithmic processing technologies expanded further these threats to caveats pertaining to dependence, fairness, due process, auditability, transparency, and knowledge asymmetries [57, 64]. In this regard, many computer and human scientists assert that the discriminatory nature of ML algorithms used by big data technologies "*prioritize information in a way that emphasizes or brings attention to certain things at the expense of others*" [35, 40]. In a recent report[6] published by the Ethics Advisory Group, a group set up by the EDPS, the interactions based on algorithmic profiling are described as exacerbating information imbalances between decision-making governments and companies on the one hand and individuals on the other hand.

However, as private and public entities worldwide rely more and more on predictive algorithmic assessments and profiling methods to make important decisions about individuals and to steer social and technological processes, the need of dealing with these threats grows dramatically [17, 27]. Therefore, we analyse below the most well known challenges arising from the big data and algorithmic processing techniques when used to profile and to make automated decisions that affect people's lives.

Biased information

While advocates of automated processing applauded the removal of human beings and their flaws from the assessment process claiming that automated systems rate all individuals in the same way, thus averting discrimination, Citron and Pasquale [27, 106], among other scholars, argued that this account is misleading because when humans program predictive algorithms, their biases and values are embedded into the software's instructions. Indeed, often profiling and automated decision-making systems mine large and diverse datasets containing inaccurate and biased information in order to create derived or inferred data about people. But when these data are inaccurate may lead to incorrect predictions and scores about their behaviour, health, creditworthiness, or insurance risk, challenging thus the general fairness of the system. Fairness in decision making systems has been first explored and formalized as a generalization of the notion of differential privacy [43] by Dwork et al. [44] who defined fairness as the extent to which similar individuals are treated similarly by the system. An example of such unfair processing is when algorithms place a low score on occupations like migratory work or low-paying service jobs, resulting thereby, even with no discriminatory intent, to unfairly impact consumers' loan application outcomes if a majority of those workers are racial minorities [27]. Hence, it is argued that in relation to the fair processing when profiling methods are used, it is important to distinguish between the concept of unintentional discrimination as classification

[6] https://edps.europa.eu/data-protection/our-work/publications/ethical-framework/ethics-advisory-group-report-2018_en.

or prioritization of information and unfair discrimination as a conscious choice that leads to prejudicial treatment [11, 73].

Algorithmic opacity

The autonomous and opaque nature of ML algorithms signifies that decisions based on their outputs may only be identified as having been discriminatory afterwards—when the impacts have already been felt by the people discriminated against [145]. As Burrell describes in [21], the opacity of ML algorithms may stem either from intentional corporate or state secrecy, or from technical reasons such as technical illiteracy or the characteristics of ML algorithms and the scale required to apply them usefully. The latter is the case when in certain algorithms the number of possible features to include in a classifier rapidly grows way beyond what can be easily grasped by a reasoning human, or when machine optimizations are employed based on training data which do not naturally accord with human semantic explanations. As she explains, this is the reason why ML is applied to the kind of problems for which encoding an explicit logic of decision-making performs very poorly [21].

Misrepresented data

Beyond the biases embedded into the systems, hidden bias may be produced when misrepresented data are fed into these systems, questioning thereby the general fairness of the processing [145]. Yet human beings, due to the complexity of the applied algorithms, cannot always properly intervene in repairing the possible original bias occurred in the data collection phase of the decision-making process [57]. This is the case reported in ProPublica's study[7, 8] where 7 000 risk scores, produced by a ML tool used in some US states to predict the future criminal behaviour of defendants, were analysed and the findings revealed discrimination based on race, with black defendants falsely classified as future criminals on nearly twice as many occasions as white defendants. Admittedly, as authors in [70] note, it is difficult to challenge inaccuracies caused by misrepresented data that are used to predict an algorithmic profile because these inaccuracies are not about the individual's actual behaviour, but rather the reported behaviour that may has been, either intentionally or unintentionally, misrepresented. Such misrepresentations may further affect people's recruitment prospects in cases where big data profiling based on candidates' choice of installed browsers is used for recruitment purposes.[9]

[7] https://www.propublica.org/article/machine-bias-risk-assessments-in-criminal-sentencing.

[8] http://www.dailymail.co.uk/sciencetech/article-3606478/Is-software-used-police-identify-suspects-racist-Algorithm-used-predict-likelihood-reoffending-biased-against-black-people-investigation-claims.html.

[9] https://www.economist.com/blogs/economist-explains/2013/04/economist-explains-how-browser-affects-job-prospects.

Correlation instead causation

Big data analytics accuracy also suffers from a logical fallacy when their results offer insights into irrelevant factors. It has been thoroughly outlined that big data processes, due to the use of the employed analysis methods which are picking up things that are not there, deal with correlation rather than causality [23, 91, 139]. Hence, the distinction between correlation and causation is very important to overcome this fallacy because correlation indicates only a probability, not a certainty. Therefore, it has been argued that organizations using ML algorithms to discover associations need to appropriately consider this distinction and the potential accuracy (or inaccuracy) of any resulting decisions [70]. Along the same lines, Diakopoulos [40] points out that "*one issue with the church of big data is its overriding faith in correlation as king. Correlations certainly do create statistical associations between data dimensions. But despite the popular adage, "correlation does not equal causation", people often misinterpret correlational associations as causal*". A good example demonstrating this fallacy is the spurious correlations project[10] where various deceptive correlations are depicted, e.g. the correlation of per capita consumption of mozzarella cheese with the civil engineering doctorates awarded.

Tyranny of minority

Beyond the aforementioned issues, Barocas and Nissenbaum have indicated [10] that profiling and automated decisions-making processes based on big data suffer from the "*tyranny of the minority*" where the willingness of few individuals to disclose information about themselves may implicate others who happen to share the same group profile with them, and particularly the same observable traits that correlate with the traits disclosed (see also 3.4.2). This is commonly referred as the case of "*creditworthiness by association*" as reported in [31]. In that report, several commenters explained that some credit card companies have lowered a customer's credit limit, not based on the customer's payment history, but rather based on analysis of other customers with a poor repayment history that had shopped at the same establishments where the customer had shopped.

Exposure of sensitive data

Problems of service discrimination or exclusion are also arising when profiling reveals sensitive personal information, such as the medical condition of a user or her propensity to develop a certain disease, out of seemingly harmless information. Therefore, although the issues of discrimination are most often associated with the impacts of the processing of personal data, Hildebrandt argues that the difference between data and personal data becomes unimportant when profiling infers highly

[10] http://www.tylervigen.com/spurious-correlations.

sensitive information out of seemingly trivial and/or anonymous data [65]. As Zarsky concisely summarizes, discrimination carried out nowadays is data driven, often does not involve intent, and is not split along the simple clear lines of the noted special categories of data [153].

6.2.5 The Risks of Tax and Financial Profiling

As discussed above, big data analytics used to infer and predict behaviours, trends, choices, and preferences, lie at the heart of modern algorithmic data processing and facilitate advanced profiling and automated decisions processes employed extensively by both the private and the public sectors. In the tax and financial domain, in particular, big data and ML algorithms have paved the way towards the bulk accumulation of tax and financial data which are exploited to either provide novel financial services to consumers or to augment authorities with automated conformance checks. Although private corporations are far ahead as far as their technological means are concerned, public administrations are closely following their best practices. For instance, while private companies are now able to monitor people's consumption patterns to predict future trends and to provide personalized advertisement, in public administration the cases of profiling citizens are increasingly emerging and automated decision algorithms are more and more employed to substitute previously human undertaken interventions and decisions.

Against this background, international and EU policies for promoting innovation and transparency in the financial and tax domain have been increased substantially over the last years. These regulatory efforts as well as the accumulation and exploitation of big financial data and their implications to people's privacy through big data profiling techniques will be discussed hereafter. Before proceeding with the discussion on the harms arising from aggressive financial and tax profiling practices, we first explore below whether the tax and financial information can be considered to belong to the special category of data characterized as sensitive under the European data protection framework.

Is tax and financial information sensitive?

In terms of their sensitivity, tax and financial information is not specified in either the DPD or the GDPR to belong to the special categories of data labelled as "sensitive"[11] requiring stricter protections than for those anticipated for other types of personal

[11] The GDPR defines as sensitive the personal data that reveal racial or ethnic origin, political opinions, religious or philosophical beliefs, or trade union membership, the genetic data and biometric data, data concerning health or data concerning a natural person's sex life or sexual orientation.

data.[12] However, as it has been already pointed out[13] [65], the regime for special categories of data is no longer adequate in the era of big data analytics because practice has shown that the same data may be sensitive in one context but not in another (particularly where data are combined). Therefore, it is becoming more and more unclear whether specific categories of data are sensitive as the use of these data may or may not be sensitive depending on each context. For instance, companies and researchers use potentially "innocent" data to make sensitive distinctions between individuals [65, 76, 111]. A notorious example is the story of a retailer shop, Target, which managed to identify pregnant customers based on their shopping habits and became a top story when predicted the pregnancy of a teenager before even her father knew about it.[14]

In the light of the above, privacy scholars have repeatedly highlighted that tax and financial data are considered to be among the most sensitive forms of personal information, as they may reveal, among others, information about income, spending and savings, employment status, person's health, marital status, lifestyle, hobbies, personal belongings, and disability status [29]. This detailed and fine-grained personal information may be used to build a concrete profile of individuals' identity, including religious and political beliefs, political alliances, and personal behaviour, thereby offering an important picture of who they are [28, 29, 127]. To this end, and given the possible criminal nature of tax evasion in some states, the WP29 asserted back in 2012 [3] that personal data linked to tax "*may be deemed as sensitive data and therefore care should be taken to afford it higher standards of data protection*".

6.2.5.1 Privacy Harms Arising from Tax and Financial Profiling

Major technology companies and intelligent public authorities already have access to a lot of online data such as search terms, blogs and social connections as well as payment transactions and tax related information. By accessing such fine grained information which eventually describes a detailed people's behaviour, companies and authorities can develop highly precise individual profiles which can be later used either for influencing the consumers' choices or for predicting future legal and tax liabilities. Still, as it is highlighted by many studies [48, 148], while many government agencies are increasingly smart about using data analytics to improve their operations and services, most agencies lag behind the best private sector firms and face challenges related to resource and infrastructure constraints as well as poor

---

[12] The US law, although extends heightened protection to certain data through specific laws and regulations, does not globally recognize types of data that receive heightened protection across various laws akin to EU-style "sensitive" data [125].

[13] https://iapp.org/news/a/gdpr-conundrums-processing-special-categories-of-data/.

[14] https://www.forbes.com/sites/kashmirhill/2012/02/16/how-target-figured-out-a-teen-girl-was-pregnant-before-her-father-did/#72a898e66686.

initial scoping.[15] These shortcomings, along with the fact that corporations do not have the same mandate for public accountability, led some scholars to argue that much of what private companies are best at doing would not be easily transferred to tax or public context in general [40, 59].

Private companies, by exploiting big data and ML capabilities, can recommend customer-specific products either for increasing their returns or for replacing existing products for new ones. However, even when profiled for marketing and advertising purposes most consumers do not like the fact of being monitored or identified. According to a case reported in [115], the Dutch bank ING was planning "*to explore if customers would be interested in receiving tailored discounts from third parties in line with their spending behaviour*", an intention that raised many negative reactions from customers and media and subsequently compelled the bank not to move forward with its plans.[16] Behavioural profiling commenced by Facebook which in 2015 changed its terms of services in order to allow the use of its customer data for commercial purposes, such as targeted advertisement, has also provoked legal actions of German, French and Dutch authorities against the firm[17,18,19] [147]. Yet, profiling practices can also become even more intrusive, like in the cases where people's credit limits is being lowered based on an analysis of the poor repayment histories of other people who shopped at the same stores as them [31] or the previously described case of the retailer who managed to predict customers' pregnancy based on the consumption of just 25 products [106]. Nevertheless, while commonly profiling based on consumer payment data can be used for harmless causes like marketing, personalized advertising and price discrimination, there are cases where can be used for more malicious ones like identity theft and social engineering [115]. Research literature is full of cases where online tools are using sophisticated algorithmic scoring techniques to target on consumers at moments when they are likely to be especially vulnerable to low-value, short-term credit products with usurious interest rates and highly unfavourable terms [71].

In the tax domain, profiling refers to the categorization of taxpayers into risk profiles based on the utilization of big data technologies where vast amount of data about them are collected through various sources [139]. In fact, tax information can be

[15] A recent study [148] of 27 public sector ML practitioners across 5 OECD countries about the faced challenges of understanding and instilling public values into their work revealed that there is a disconnection between institutional realities and research outcomes toward transparent and non-discriminative ML systems. Researchers concluded that for transferring the values of fair and accountable ML into public sector, the respective processes should be studied in vivo, in the messy, socio-technical contexts in which they inevitably exist since issues like fairness have been shown to come with technically difficult to reconcile, or even irreconcilable, trade-offs—or concerns raised that explanation facilities might work better for some outputs than for others.

[16] https://www.ing.com/About-us/ING-and-the-use-of-customer-data.htm.

[17] https://autoriteitpersoonsgegevens.nl/en/news/dutch-data-protection-authority-facebook-violates-privacy-law.

[18] https://www.theguardian.com/technology/2017/may/16/facebook-facing-privacy-actions-across-europe-as-france-fines-firm-150k.

[19] https://www.theguardian.com/technology/2018/feb/12/facebook-personal-data-privacy-settings-ruled-illegal-german-court.

cross-indexed by the public revenue authorities against other digital personal information maintained by domestic and foreign governments (e.g., customs, criminal or immigration data) or by the private sector (e.g., records of consumer purchases) to allow for a detailed profile of an individual to be put together from formerly discrete bodies of data. This detailed profile can be used for purposes outside of traditional tax concerns such as a part of an investigation for terrorist financing schemes [28]. In the US, the Inland Revenue Service (IRS) is entitled to collect an enormous amount of private information, such as sleeping habits, individuals' hobbies, reading preferences (where and for how long a taxpayer's gaze falls on certain screens), religious affiliations, travel plans, medical conditions, weight and doctor's recommendations about it, to name a few [59, 143]. This is why the IRS has been described by its Commissioner[20] as "*an information intensive enterprise*" which works on "*the organization of data and ultimately the knowledge and intelligence we extract from the information*". Yet, it has been argued that there are some IRS methods, mostly unknown to the general public, which violate fair information practices. According to Houser and Sanders [70], the IRS is reported to have used automated computer programs (known as spiders) and big data analytics to sort through and mine social media sites[21] not only about a taxpayer who is being audited but even for potential tax violators not selected for audit.

Likewise, in the UK the integration of predictive analytics tool with big data warehouses has built the "*all-seeing eye*" of the HMRC (Her Majesty Revenue and Customs) that targets taxpayers' online information and enables drilling down into over one billion pieces of data, analysing the digital patterns of behaviour, payments and money flows of individuals and businesses.[22] HMRC tools interrogate 30 databases in total, containing not only information spontaneously available in government departments but data also found online as well, like on the Airbnb and the e-bay,[23] in order to apply sophisticated profiling and modelling techniques and to search for patterns and behaviours that signify tax anomalies.[24,25] It also scrutinize the digital footprint that people leave when they use the Internet, searching social media for holidays and luxury items information which then is used to build a lifestyle profile of individuals who are under investigation for tax or benefits fraud.[26] HMRC, utilizing the well-known industry model of "understanding customers' behaviour",[27] is further requesting bulk data from third parties, like insurance companies and hospitals or payments to general practitioners and dentists, when there is evidence of

[20] https://www.irs.gov/newsroom/commissioner-doug-shulman-speaks-at-aicpa-meeting.

[21] http://washington.cbslocal.com/2014/04/16/report-irs-data-mining-facebook-twitter-instagram-and-other-social-media-sites/.

[22] https://www.computing.co.uk/ctg/feature/2244719/connecting-the-dots-at-hmrc.

[23] https://www.telegraph.co.uk/tax/return/taxman-unleashes-snooper-computer-information-does-have/.

[24] https://perma.cc/F33W-M9FL.

[25] https://www.accountancylive.com/hmrcs-connect-targets-taxpayers-online-information.

[26] https://www.ft.com/content/0640f6ac-5ce9-11e7-9bc8-8055f264aa8b.

[27] https://www.capgemini.com/wp-content/uploads/2017/07/ss_hmrc_adept.pdf.

widespread tax evasion or under-reporting. Within this scope of exchanging data with other agencies, the recently enacted Digital Economy Act 2017[28]—which regulates matters of information sharing between public bodies in respect of Public Service Delivery, Debt and Fraud—allows the HMRC and other public sector authorities to exchange and disclose personal data for the prevention of fraud and the recovery of debts. Nevertheless, while the Digital Economy Act 2017 maintains safeguards on privacy and anticipates various provisions and codes of practice relating to the confidentiality of personal information,[29] it has been also widely criticized for excessive disclosure risks as it provides a number of exemptions that permit the disclosure of confidential information.[30,31,32]

6.2.5.2 Policies for Tax and Financial Transparency

The number of international and EU policies towards collecting and exchanging large amount of personal tax and financial data to facilitate innovation and to promote transparency in the financial and tax domain has been increased substantially over the last years. Accordingly, there is an escalating dominance of the principle of transparency for personal financial and tax information in the expense of privacy. In fact, following the 2008 global financial crisis numerous international policy makers argued that the need for free and unfettered access to personal financial and tax data in order to combat criminals and terrorists supersedes the principle of a right to privacy[33] [127]. In this respect, transparency in the public domain has been presented as the solution and privacy as an obstacle to policy success [60, 127]. Similarly, in the private sector privacy policies and regulations have been commonly linked to inhibiting innovation and directly affecting the economic growth and the efficacy of emerging technologies [48, 52].

In this perspective, tax authorities are currently engaged in automatic ways of exchanging information for combating tax evasion and fraud, while at the same time private corporations are exploiting innovative ways to hold and capitalize on personal financial information; all within the EU and the international regulatory framework. The Organization for Economic Co-operation and Development (OECD) and the EU have been engaged in an unprecedented effort towards the Automatic Exchange of Information (AEOI) of tax related data among their jurisdictions whereas the recent EU legislations on open banking and financial services are challenging the

[28] http://www.legislation.gov.uk/ukpga/2017/30/contents/enacted.

[29] https://www.gov.uk/government/publications/digital-economy-act-2017-part-5-codes-of-practice.

[30] https://www.lexology.com/library/detail.aspx?g=f137a29a-4145-4bf9-bd49-ae9e4c77a1fc.

[31] https://www.twobirds.com/en/news/articles/2017/uk/very-latest-data-protection-changes.

[32] https://www.theguardian.com/commentisfree/2017/feb/05/the-guardian-view-on-the-digital-economy-bill-a-last-chance-to-get-it-right.

[33] This tendency is very often justified on the grounds of a "nothing to hide, nothing to fear" logic arguing that only the guilty have secrets to hide, an argument which Solove efficiently confronts in [131].

status quo of the traditional bank sector. The catalyst for the worldwide expansion of the AEOI was a piece of legislation adopted in the US in 2010 called the Foreign Account Tax Compliance Act (FATCA) [38, 50, 141]. FATCA's enactment resulted in the OECD's and the EU's ground-breaking efforts towards common reporting and due diligence standards to support a global system for combating offshore tax evasion [68]. The outcomes of these efforts were the OECD Common Reporting Standard (CRS) for automatic worldwide exchange of financial account information and the EU Directives on Administrative Cooperation (DACs) which introduced broad exchange of information within the EU without prior request [123]. For more information on these policies please refer to [113].

In the private sector, two of the most prominent legislations that influence the landscape of banking and payment services are the 2nd Payment Services Directive (PSD2) and the 2nd Markets in Financial Instruments Directive (MiFID2). PSD2 aims to increase the pan-European competition and participation in the payments industry also from non-financial institutions and to promote the development and use of innovative online and mobile payments, such as through open banking. MiFID2 aims to strengthen investor protection and to improve the functioning of financial markets in a more efficient, resilient, fairest and transparent way possible. For more information on these directives please refer to our work in [113].

6.2.5.3 Impact on Profiling and Privacy

Although ultimately the purpose of the aforementioned initiatives is to promote financial and tax transparency as well as innovation and new individualized services by exploiting "big" tax and financial data, their impact on individuals' privacy through the facilitation of building profiles of consumers and taxpayers based on payment and tax related data is overwhelming. As a matter of fact, the vast collection and utilization of "big" tax and financial data raise major considerations around privacy and data protection rights, especially when these data are fed to clever algorithms in order to build detailed personal profiles or to take automated decisions which may exceptionally affect people's lives.

Unavoidably, opening up and sharing banking transactions data under the PSD2, it will provide a huge amount of payment data to companies which, by knowing the spending behaviour of individuals, they will be able not only to analyse the data and guide them to better decisions regarding their money spending[34] but also to construct a full spending and consuming profile. Since electronic payments, unlike cash, link a particular person with a particular purchase, the monitoring of consumption patterns, as well as the tracking of a person's movements becomes possible. As demonstrated earlier, people's spending patterns comprise a valuable information precisely because it is possible to extrapolate inferences about the individuals in question as payment data are the exact image of their behaviour, choices and preferences, all so far considered as private [127].

---

[34] https://www.theguardian.com/money/2018/jan/08/open-banking-bank.

Under the new rules imposed by the PSD2, the ownership of these data will be essentially transferred to the consumer, meaning that account holders will be able to give companies, other than their own bank, permission to access their details. Obviously this granting should be accomplished in an easy and secure fashion. Therefore, strong customer authentications, like two factor authentication, are specified under the RTS as a way of ensuring that data can be shared securely. Two factor authentication specifies that, apart from using the first knowledge factor (e.g. PIN) for accessing a service, a second factor based on either possession (e.g. token) or inherence (e.g. biometrics) is needed as well. Yet, the use of inherence as the second factor to cater for both the security requirements and user experience priorities of PSPs, it enables the incremental collection of big biometric data that can be later used for profiling inferences. Apart from biometrics, which is already in widespread use, another important subset of inherence is behavioural profiling. By assessing the customer's location and behaviour against their usual patterns, corporations can gain a clearer view of the risks and the level of authentication required. Even though behavioural profiling is a comparatively new mechanism that is currently being used by the industry as an augmentation to strengthen fraud controls,[35] its future contribution to the construction of an integrated individual profile when combined with other personal data is indisputable.

The AEOI framework under which governments automatically exchange cross-border big data consisting of bulk taxpayer information to combat international tax evasion and better target audits of aggressive international tax planning [29], while being potentially revolutionary [139], it has been also heavily criticized, on the one hand, due to its obligations of exchanging not only information relating to a single taxpayer (or a specific group of taxpayers) but of a bulk information without any indications of non-compliant behaviour of the taxpayers[36] [7, 13, 39, 118, 132]; on the other hand, due to the removing of several existing privacy safeguards to improve the efficiency of the exchange process [8, 41, 68, 118]. Above all, however, it has been accused as facilitating the construction of detailed taxpayers profiles that may be used for purposes beyond tax context. For instance, aiming at fighting offshore tax fraud, tax authorities are inclined to use phone records which may reveal whether an individual is contacting an offshore service provider based in a tax haven[37] [29]. In that respect, AEOI along with national laws for data retention such as DRIPA, which is currently under revision,[38] and for data sharing such as the Digital Economy Act 2017, may impact hugely on citizens' privacy.

[35] https://www.accenture.com/_acnmedia/PDF-40/Accenture-PSD2-Open-Banking-Security-Fraud-Impacts.pdf.

[36] http://freedomandprosperity.org/2017/blog/new-tax-oppression-index-shows-grim-toll-of-oecds-statist-agenda/.

[37] Because residents, in order to avoid paper trails when set up offshore trusts, proceed to oral instructions regarding disbursements.

[38] https://www.computerworlduk.com/security/draft-investigatory-powers-bill-what-you-need-know-3629116/.

The interested reader may refer to our extensive discussion on the AEOI framework in [113] where we also thoroughly analyse the challenges of the data protection rights under these regulatory initiatives.

6.2.6 Towards Accountable, Transparent and Fairer Profiling and Automated Decision Making

As previously discussed, the practices of profiling and automated decision making may provide a fertile ground for discriminating and unfair processing of individuals and groups. Nevertheless, in the era of AI and ML the accountability of algorithmically automated decision systems occupies increasingly the interest of legal and technical research communities who call for automated decisions to be accountable to the public and individuals to have the right to inspect, correct, and dispute inaccurate data, to know their sources, or, at the very least, to have a meaningful form of notice and a chance to challenge predictive decisions that harm their ability to obtain credit, jobs, or other important opportunities [21, 27, 40, 117, 124, 152]. Yet according to some scholars, detecting discriminatory decisions in hindsight is not sufficient and hence they urge big data analysts to find ways to build discrimination detection into their systems to prevent such decisions being made in the first place. This can be achieved by introducing and implementing appropriate algorithmic tools and interventions to both identify and rectify cases of unwanted bias so as, apart from making more accurate predictions, to offer increased transparency and fairness as well [53, 145]. Diakopoulos [40] on the other hand explains that transparency, as a medium that facilitates accountability, should be demanded from the government and should be exhorted from the industry, whereas Zarsky [152] identifies three stages in which the transparency requirements should be met: in the data collection stage, in the data analysis stage, and in the final ex post usage stage of the produced decision. While many scholars have pointed out that obviously in the collection stage legitimate arguments for some level of big data secrecy commonly related to corporate intellectual property and national security secrets may be raised [27, 117, 152], Richards and King [117] identify this secrecy as a big data "*transparency paradox*" since even though big data promises to make the world more transparent, its collection is invisible, and its tools and techniques are opaque.

Pursuing transparency, Danielle Citron [26] called more than a decade ago for a "technological due process" in the automated decisions context to underline that these decisions cannot be made within black boxes but due processes are needed to entail limits on fine-grained personalization in a range of public administrative processes [30, 59]. In the big data context, these "due processes" should apply to both government and corporate decisions derived from big data analytics, and when these decisions affect individuals those people should have a right to know on what basis those decisions were made [26]. In public administration in particular, there are additional ethical, social and legal constraints that probably will rule out a range of "private sector-like" uses of predictive modelling, profiling and algorithmically automated decisions practices on similar targeted public services, e.g. for tax or health relief [48]. For instance, while private companies aim to monitor, predict, and change consumer behaviour, and hence their analysis does not require legal-standard accuracy, when profiling and automated decision-making is used in the tax or other public-related context, the results must interpret the law and would need to analyse consequences within legal standards of accuracy since any errors may violate citizens legal rights. Furthermore, as Hatfield notes [59], any system to "automate" tax or other legal decision-making would be tremendously complex as it would have to reveal how the decision was made and how the legal values were interpreted and applied in a way that the taxpayer could understand and respond. In that respect, Cohen [30] proposes the concept of "semantic discontinuity", as opposed to "seamless continuity", "*as a function of interstitial complexity within the institutional and technical frameworks that define information rights and obligations and establish protocols for information collection, storage, processing, and exchange*". She also notes that "semantic discontinuity" can be conceptualized more generally as a right to prevent precisely targeted individualization and continuous modulation, and serves similar ends as what the legislators of the GDPR are indented to deal with when introducing the "Right to be Forgotten" [112].

A well known application of introducing transparency and accountability in public administration domain is found at the city council of New York which in 2017 introduced a bill[39] that would require the city to make public the up to then invisibly used algorithms in all kinds of government decision-making systems used for detecting and addressing financial fraud, crimes, as well as public safety and quality of life issues.[40,41] Supported by many transparency and privacy advocates along with social and computer scientists, the bill passed, albeit amended.[42,43] The amendment foresees for an experts task force to be created in order to review city agencies' use of algorithms and respective policies and to develop a set of recommendations on a range of issues, including which types of algorithms should be regulated, how citizens can meaningfully assess the algorithms' functions and gain an explanation of decisions that affect them personally, and how the government can address cases in which a person is harmed by algorithmic bias.[44] Although far behind of what the original bill anticipated, the amendment is considered to have a significant impact on the automated decision-making by public authorities. As of May 2018, the Automated Decision Systems Task Force, the first of its kind in the US, was announced

[39] https://www.nytimes.com/2017/08/24/nyregion/showing-the-algorithms-behind-new-york-city-services.html.

[40] http://www1.nyc.gov/site/analytics/index.page.

[41] https://www.oreilly.com/ideas/predictive-data-analytics-big-data-nyc.

[42] https://laws.council.nyc.gov/legislation/int-1696-2017/.

[43] https://www.aclu.org/blog/privacy-technology/surveillance-technologies/new-york-city-takes-algorithmic-discrimination.

[44] https://www.newyorker.com/tech/elements/new-york-citys-bold-flawed-attempt-to-make-algorithms-accountable.

with the task to develop a process for reviewing New York City's algorithms and automated decision systems through the lens of equity, fairness and accountability.[45] In August 2018 several experts in the field of civil rights and AI co-signed a letter to the Task Force providing recommendations such as creating a publicly accessible list of all the automated decision systems in use, consulting with experts before adopting an automated decision system, and creating a permanent government body to oversee the procurement and regulation of automated decision systems.[46,47] The Task Force were to produce its first report recommending procedures for reviewing and assessing City algorithmic tools in December 2019. However, the long anticipated report[48] has been widely criticized for not meeting the expectations of interpreting how City's AI algorithms really work since the Task Forced was never fed with the necessary information by the City's officials.[49,50,51]

Despite the above, transparency has been critically seen by many scholars as an inadequate measure for accountability of modern algorithmic systems [2, 77]. Kroll et al. [77] have thoroughly demonstrated that transparency is not enough and not even possible in automated decision making systems based on ML algorithms. In this regard, they introduce computational methods that can provide accountability for procedural regularity even when some information is kept secret. These methods can be used alongside transparency and auditing and can be applied to all computer systems [77]. In addition, interpretability, that is providing ex ante and ex post explanations on the inferred decisions as a mean of accountability, holds the attention of a big part of scientific and legal community in terms of its effectiveness in the algorithmically decision supported systems and its benefits compared to its cost [42]. As Lipton analyses in [85], the term interpretability does not refer to a monolithic concept but it can be addressed within the context of various model properties and techniques. On the other hand, however, Hildebrandt remarks [67] that explanation, as a notion of interpretability, in itself does not imply justification since a decision of an automated system should be justifiable independently of how the system came to its conclusion.

Depending on the stage of which the lack of transparency and accountability may be identified in a particular algorithmic application, different course of actions, ranging from legislative, to organizational and technical, are likely to mitigate its problems [21]. While the research work in the technical layer is booming as various

[45] https://www1.nyc.gov/office-of-the-mayor/news/251-18/mayor-de-blasio-first-in-nation-task-force-examine-automated-decision-systems-used-by.

[46] https://ny.curbed.com/2018/8/24/17775290/new-york-city-automated-decision-systems.

[47] http://assets.ctfassets.net/8wprhhvnpfc0/1T0KpNv3U0EKAcQKseIsqA/52fee9a932837948e3698a658d6a8d50/NYC_ADS_Task_Force_Recs_Letter.pdf.

[48] https://www1.nyc.gov/assets/adstaskforce/downloads/pdf/ADS-Report-11192019.pdf.

[49] https://www.theverge.com/2019/4/15/18309437/new-york-city-accountability-task-force-law-algorithm-transparency-automation.

[50] https://www.citylab.com/equity/2019/12/ai-technology-computer-algorithm-cities-automated-systems/603349/.

[51] https://redtailmedia.org/2019/04/08/members-and-watchdogs-of-nycs-ai-task-force-cry-transparency-foul/.

techniques and methods for interpretable ML algorithms which provide explanations on the derived profiling classifications and decisions have been proposed [79, 80, 116, 150], in the legislative layer the advancements are quite reserved. It has been suggested that automated decision-making systems should be subject to licensing and audit requirements when they enter critical settings like employment, insurance, and healthcare [27, 40], whereas other scholars proposed for an oversight board or a federal agency to ensure that algorithms produce accurate, fair and effective decisions [70, 144]. The idea of regulators to be able to test automated decisions-making systems to ensure their fairness and accuracy had also presented by Citron and Pasquale who argued that individuals should be granted meaningful opportunities to challenge adverse decisions based on scores or decisions miscategorizing them [27]. In this respect, proposals have been made toward regulations that compel information with at least, and always depending on the context of each algorithm, five broad categories of information: human involvement, data, model, inference, and algorithmic presence [40].

By all means, the EU GDPR regulation has been seen as an attempt towards minimizing the unwanted privacy implications of ubiquitous and pervasive mobile computing and ML big data processing, as well as a way of mitigating the risks of aggressive profiling and unaccountable automated decision-making. Therefore, in what follows we analyse how the GDPR could achieve these goals in the big data and ML era.

6.3 Mitigating Privacy Risks Under the GDPR

In view of the afore-mentioned privacy risks, the concept of building trust is constantly under stake when most users today are not even aware of the data processing procedures undertaken by businesses and authorities with their personal data. As Spiekermann et al. underline in [134]: "*If they* [users] *learned about today's volume and business done with their data among third parties, they may be surprised and feel betrayed. No matter whether and to what extent first party companies have engaged in data deals themselves, they could all be hit by a backlash from users once they find out*". This adverse reaction has also pointed out by Mittelstadt et al. [96] who argue that "*the tension between personal big data and privacy often triggers a "whiplash effect", by which overly restrictive measures (especially legislation and policies) are proposed in reaction to perceived harms, which overreact in order to re-establish the primacy of threatened values, such as privacy*". For many big data enthusiasts and privacy skeptics, the GDPR constitutes an emergent "backlash", an overwhelming reaction from regulators to the bursting exploitation of personal data dominating not only the way industry performs business but academia conducts research as well.

In fact, as privacy often contradicts modern research and industry practices, privacy concerns have split a large share of academia. To mitigate the harms from processing personal data and the consequences in individual privacy many workable methods and technical solutions have been introduced over the last decades:

k-anonymity [137], l-diversity [87], t-closeness [81], differential privacy [43], data aggregation [83], and data obfuscation [9], among others. While each of these methods may be appropriate to per case approach, the general concepts underpinning modern privacy aware systems are based on the *"privacy by design"* principles which are anticipated explicitly by the GDPR (see 3.3). However, big data characteristics by their very nature go against these principles [45, 96]. Under data minimization and purpose limitation organizations are required to limit the collection of personal data to the minimum extent necessary to obtain their legitimate goals and to delete data that is no longer used for the purposes for which they were collected. On the contrary, big data business model encourages collection of more data for longer periods of time [142].

In view of the above, a large proportion of research community urges for loosened privacy regulation and increased trust on the research ethics arguing that the fact researchers can identify individuals and all of their actions is a necessary trade-off for high quality research [37], whereas others argue that the regulation does not fully grapple with the challenges posed by big data and a way forward would be the experimentation with a more flexible approach to regulation through the creative use of codes of conduct [120]. All these arguments against strict privacy regulations are based on the inevitably reality that data utility decreases when privacy increases, a fact that urges data driven business world to warn against the overly broad regulatory definitions of personal data and to highlight that regulations on data protection and privacy may preclude economic and societal benefits [61]. Stated otherwise by Ohm, "*no useful database can ever be perfectly anonymous*" [101]. Notwithstanding this clash, most scholars agree that there cannot exist big data without privacy since the protection of personal data is, first of all, in the interest of the big data analytics service providers who will ultimately have to cope with this challenge [36].

The GDPR, taking into account both risks and challenges that big data may bring upon citizens, introduced the new legal term of pseudonymization (Article 4(5)) in order to describe data that could be attributed to a natural person by the use of additional information, which must be kept separately and be subject to technical and organizational measures to ensure non-attribution. While the use of pseudonymization is encouraged in many occasions, pseudonymized information is still considered a form of personal data and hence, a value to protect.

In what follows, we investigate the protection of individuals against the risks of profiling and automated decision-making in the big data era and in the light of the GDPR. In this respect, we delve into the regulatory EU efforts towards fairer and accountable profiling and automated decision processes, and in particular we examine the extent to which the GDPR provisions establish a protection regime for individuals against advanced profiling practices and discriminatory automated decisions, enabling thus accountability and transparency. Moreover, we discuss and propose implementation scenarios for addressing, even partially, the raised concerns.

6.3.1 *Profiling and Automated Decision Making Under the GDPR*

As Hildebrandt noted back in 2008 , for a long time, profiles, as opposed to personal data, didn't have a clear legal status and therefore the protection against profiling was very limited [64]. In 2010, the Council of Europe published its recommendation on "*the protection of individuals with regard to automatic processing of personal data in the context of profiling*" [32]. Therein, the notion of profile was defined as "*a set of data characterizing a category of individuals that is intended to be applied to an individual*" and the profiling was referring to "*an automatic data processing technique that consists of applying a "profile" to an individual, particularly in order to take decisions concerning her or him or for analysing or predicting her or his personal preferences, behaviours and attitudes*". The GDPR, largely inspired by this definition, provides a similar term for profiling: *"profiling" means any form of automated processing of personal data consisting of the use of personal data to evaluate certain personal aspects relating to a natural person, in particular to analyse or predict aspects concerning that natural person's performance at work, economic situation, health, personal preferences, interests, reliability, behaviour, location or movements"@@@ (*Article 4*).* Yet, according to WP29 guidelines on the automated individual decision-making and profiling under the GDPR [5], the two definitions are not identical to the fact the recommendation excludes processing that does not include inference.

The GDPR in Article 22 specifically provides for people's right not to be subject to a decision based solely on automated processing, including profiling, if this profiling "*significantly affects*" them. While a corresponding definition in Article 15[52] of the DPD had been criticized by scholars as providing limited protection against application issues of profiling [22, 124], the scope of the GDPR Article 22 is much broader in terms of the rights of the data subjects when their personal data are being processed for profiling purposes [74, 94]. Although the choice of the term "*right*" in the provision suggests that the Article applies when it is actively invoked by the data subject, the WP29 guidelines clarify that the article "*establishes a general prohibition for decision-making based solely on automated processing. This prohibition applies whether or not the data subject takes an action regarding the processing of their personal data*" [5]. In other words, this prohibition, which is also suggested by the text "*should be allowed where expressly authorized*" in recital 71, implies that processing under Article 22(1) is not allowed generally and hence individuals are automatically protected from the potential effects this type of processing may have. Therefore, this right cannot be considered to be a special form of opt-out as it has been claimed thus far [88, 94]. Certainly, this general prohibition is legitimate unless one of the exceptions of Article 22(2) applies, that is when the automated decision making is necessary for the performance of or entering into a contract; or is authorized by Union

[52] DPD Article 15 includes "*automated processing of data intended to evaluate certain personal aspects relating to him, such as his performance at work, creditworthiness, reliability, conduct, etc.*".

or Member State law to which the controller is subject and which also lays down suitable measures to safeguard the data subject's rights and freedoms and legitimate interests; or is based on the data subject's explicit consent. Moreover, as WP29 emphasizes, the Article 22(1) prohibition *only* applies in specific circumstances when a decision based solely on automated processing, including profiling, has a legal effect on or similarly significantly affects someone.[53] While it has been suggested that this right can be circumvented relatively easily by inserting even nominal involvement of a human in the loop [149, 153] (as the provisions is restricted to"solely" automated processing), the GDPR text as well the WP29 guidance identify that there are still many situations where the right is very likely to apply, such as credit applications, recruitment and insurance [145]. Nevertheless, WP29 notes that "*targeted advertising based on profiling will not have a similarly significant effect on individuals*", raising thus concerns around cases where targeted advertising relies on highly intrusive profiling based on behavioural observed, inferred or predicted data [72, 106].

Furthermore, Article 22(3) specifies that "*the data controller shall implement suitable measures to safeguard the data subject's rights and freedoms and legitimate interests, at least the right to obtain human intervention on the part of the controller, to express his or her point of view and to contest the decision*". Apart from the fact that the Article does not elaborate on what these safeguards are, beyond "*the right to obtain human intervention*", it has been pointed out that the wording indicates that in the absence of decision-making, profiling alone does not give rise to safeguards under Article 22 [72]. Yet, the GDPR still gives rise to safeguards under Articles 13 to 15 to provide information on the processing. Actually, for many scholars the novelty of the GDPR profiling provisions is contained in Articles 13, 14 and 15 which oblige data controllers to provide "*information as to the existence of automated decision-making, including profiling*", and "*meaningful information about the logic involved, as well as the significance and the envisaged consequences of such processing for the data subject*" [17, 153]. Given this wording and the ML algorithmic opacity, scholars have unavoidably prompted the question what is required for data controllers to provide a meaningful information to explain not only an algorithm's decision but also the envisaged consequences of its processing [21, 53, 73].

As a matter of fact, the discussions on whether the GDPR implements an *ex ante* or an *ex post* right to explanation as a way to achieve accountability and transparency in automated decision-making provoked a heated debate among the legal, privacy and ML community, with some scholars arguing that the GDPR does not, in its current form, implements an *ex post* right to explanation [149], while others arguing otherwise [53, 74, 94, 126]. Under a third perspective, it has been asserted that a right to an explanation in the GDPR, even if exists, it is unlikely to present a complete remedy to algorithmic harms. Instead, a right to appeal to a machine against a decision made by a human may be proved to be the more effective remedy [46, 73].

---

[53] However, there are exceptions to these circumstances, such as when the profiling activities are necessary for a contract between the data subject and the data controller, or when the profiling is authorized by Member State law to which the controller is subject, including for fraud and tax-evasion monitoring, or data subjects have given explicit consent.

Considering these arguments, the WP29 in its guidelines [5] underline that the GDPR does not require the controller to provide a complex explanation of the algorithms used or disclosure of the full algorithm. Instead, the controller should find simple ways to inform the data subject about the rationale behind, or the criteria relied on in reaching the decision. On top, given that the controller should provide the data subject with information about the *envisaged consequences* of the processing, rather than an explanation of a *particular* decision, the WP29 affirms that information must be provided about the *intended or future* processing and should include general information (notably, on factors taken into account for the decision-making process) useful for challenging the decision. While this reading clarifies that the GDPR specifies a right to an *ex ante* explanation, still it has been argued that the requirement for data subjects to be provided with "*knowledge of the reasoning underlying data processing*" in the context of decisions taken on the basis of big data-type processing is both unrealistic and deeply paradoxical, especially when they involve self-learning algorithms [119]. In this respect, counterfactual explanations have been proposed as a solution that bypasses the current technical limitations of interpretability and strikes the balance between transparency and the rights and freedoms of individuals [150].

The WP29 in its guidelines [5] clarifies further that decisions that are not solely automated might also include profiling whereas highlights the distinctions between profiling and automated decisions [5]: "*Automated decisions can be made with or without profiling; profiling can take place without making automated decisions. However, profiling and automated decision-making are not necessarily separate activities. Something that starts off as a simple automated decision-making process could become one based on profiling, depending upon how the data is used*". It also provides a couple of interesting explanations around the wide disputes provoked by Article's 22 interpretations. Taking into account that profiling practices can create a special category of "sensitive" data by inference from data which are not "sensitive" but become so when combined with other data, as well as the fact that a profile that relates to an individual and makes her identifiable is considered a type of personal data and ought to be protected [46, 57], the WP29 concludes that the rights to rectification and to be forgotten (article 16 and 17 respectively) apply both to the "input personal data" (the personal data used to create the profile) and the "output data" (the profile itself or the "score" assigned to the person) [5]. However, the precedent WP29 guidance on the right to data portability (article 20) [4] specifies that the right does not cover inferences from personal data analysis, like algorithmically or statistically derived categorizations or personalization profiles,[54] implying thereby that the inferences of a system "belong" to the system that has generated them and not to the users whose personal data feeds this system [46]. Taking further into consideration the complementary nature of the right of data portability and the

[54] In [4] WP29 specifies that "*any personal data which have been created by the data controller as part of the data processing, e.g. by a personalization or recommendation process, by user categorization or profiling are data which are derived or inferred from the personal data provided by the data subject, and are not covered by the right to data portability*".

RtbF, along with the fact that, as explained in [112], the GDPR explicitly specifies that when the exercise of the RtbF is based on the withdrawal of a previously given consent then the revocation is not retroactive, meaning that it does not apply for the processing that had taken place before withdrawal (Article 7(3)), it is deduced that profiles constructed and decisions previously taken on the basis of this information can therefore not be simply annulled.[55] In our opinion, this conclusion, supported also by the guidelines on the right to data portability as mentioned above [4], clearly contradicts the WP29 guidance on applying the RtbF on "output data", and hence it creates a serious loophole [46, 146].

As the GDPR foresees and regulates the core feature of big data analytics, namely the ability to profile individuals and to make automated decisions about them when algorithms are applied to large amounts of granular data [145], it has been forcefully argued that GDPR's implementation impact on big data practices would be substantial and highly problematic, albeit not prohibitive [92, 153]. For big data enthusiasts, the prohibition defined under the Article 22 is perhaps the most salient example of the GDPR's rejection of the big data revolution and it is actually the main reason why its predecessor, Article 15 of the DPD, was either rarely applied or even a dead letter in some MSs [153]. And given the fact that the GDPR provides persons with stricter protections from such decision making processes than its predecessor did, there have been even greater doubts as to whether it will have a significant practical impact on automated profiling decisional systems that are extremely complex and opaque [73, 94]. Therefore, some legal scholars argue that the generic key principles and procedural rights of individuals, already established under the EU data protection law since its inception, are more potent to mitigate the long-term risks of big data and algorithmic decision-making compared to the specialized provisions on automated decision-making and profiling in the GDPR [104]. Others scholars however, such as Hildebrandt, feel confident that the GDPR might allow citizens to "*have their cake and eat it too*" as they will benefit from enhanced data protection, while enjoying the innovations advanced data analytics bring about [66]. Yet, as we already discussed in 3.4.2, the fact that the GDPR applies to the profiling of individual data subjects and not of groups (since data that do not pertain to natural persons are beyond the scope of the GDPR) raises many questions on how data subjects are protected against decisions that have significant effects on them and subsequently affect their lives but they are based on group profiling [46, 73, 104, 140].

Acknowledging the aforementioned limitation of the GDPR as far as the regulation of personal data processing in the era of big data profiling analytics and automated decision is concerned, the Council of Europe published, almost a year after the GDPR's adoption by the EU, its Convention 108 Guidelines "*on the protection of individuals with regard to the processing of personal data in a world of Big Data*" [33]. According to Mantelero [90], the guidelines move away from the

[55] Still, the GDPR does not clarify what happens when the erasure is not based on the consent revocation but on some other available ground defined in Article 17(1). For these cases, it remains unclear whether a data controller is obliged to stop using the model or to go back and retrain the model either without including the erased data or even not to do anything at all.

EU traditional view in data protection regulation and provide for a more transparent approach towards the use of algorithms in decision making processes as well as extended data subjects protections against the so-called dictatorship of data. Overall, the case of big data regulation under either the GDPR or the Convention 108 Guidelines, brings forward the discussion for the future of data protection regulation in Europe, and contributes to the pursuit of regulating big data technology and specifically invasive and discriminatory profiling and automated decision practices.

6.3.2 *Implementation Challenges and Countermeasures*

While the skepticism around the reasoning of automated decision systems, the discrimination of ML algorithms, and the privacy invasiveness of big data mining still stands, eventually all these advanced technologies are allowed in many occasions due to their effectiveness. For instance, tax revenue authorities can employ such techniques to provide their services towards a fairer tax system. Yet, the major question that arises is whether the same goals can be achieved with less privacy-invasive techniques that respect the GDPR legal framework.

Currently, there is a lot of effort on the integration of privacy enhancing technologies. By leveraging csryptographic primitives such as secure multi-party computation (SMC), order preserving encryption, functional encryption, and homomorphic encryption one can perform a wide range of privacy-preserving queries along with the training and inference of ML algorithms. However, the major challenge comes from the heterogeneity and sparsity of the data since in most scenarios, the goal is not to determine whether an individual belongs in some lists, but e.g. whether her aggregated deposits or expenses from all her bank accounts are beyond a threshold. Although for this kind of privacy preserving aggregation there are several cryptographic solutions [49, 78, 107], they do not support threshold, but exact sums, and they require all parties to be simultaneously online. Similarly, range queries over encrypted data [15, 84] have as a prerequisite the use of a common public key.

Based on the above and on the fact that SMC only supports computations on data encrypted under the same public key, the introduction of an independent semi-trusted third party is needed to facilitate the requested task. In this regard, this entity would act as a broker/intermediator to allow the orchestration of collection of the financial data of individuals in an encrypted form, under a single key, and it will perform the requested operations of the authorities in an encrypted setup, thereby drastically decreasing the privacy invasive methods used so far. Since the data are encrypted, this entity cannot extract sensitive information of individuals or even differentiate them. Therefore, this entity does not need to be fully trusted.

A lot of research has been carried out recently in the field of privacy-preserving ML [19, 62, 86, 97, 121, 129]. The recent work of Li et al. [82] is aligned with the application scenario that we are dealing with in the sense that we have multiple data providers and that the analyst at the end of the protocol performs ML over the joint dataset which contains the data of individuals with minor errors. This scenario

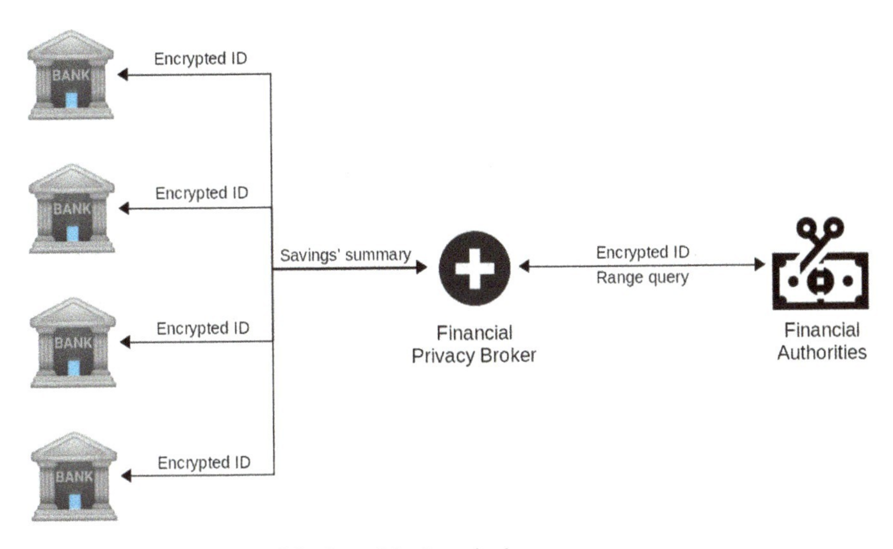

Fig. 6.2 Example use case of the financial privacy broker

provides privacy for individuals and it does not disclose the operations to the data providers. The seminal work of Graepel et al. [55], despite its inherent limitations such as allowing only two trivial classifiers, showed that one can train a ML algorithm using encrypted data. This work provided the basis for many other research studies that improve its efficiency and include far more ML algorithms [16, 18, 51, 103]. Therefore, while the ML algorithms in the privacy-preserving model may not be very efficient as their non-private counterparts, they can at least adhere to the privacy regulations of GDPR and provide good, yet not so fine-grained, results. Again, the introduction of a semi-trusted entity could significantly improve both efficiency and results.

In what follows, we call this semi-trusted entity Financial Privacy Broker (FPB). As already discussed, the main role of the FPB is to collect encrypted data from specific data sources and to provide the authorities with a range result of their aggregated data. A typical example of how we envision the operation of the FPB is illustrated in 6.2. Let us assume that financial authorities (FA) of country A want to determine whether the savings of citizen C are in the range of [m,n]. To this end, FA have to contact all cooperating banks B_1, B_2,…,B_k and request the savings of C and classify C to the corresponding class. Rather than doing this, FA send the request to FPB who will send the query to B_1, B_2,…,B_k. On receiving this, B_1, B_2,...,B_k start the two round protocol of Kursawe et al. [78], or for more efficiency the Patsakis et al. [107] (if more summaries have to be performed), and compute the aggregated summary S of C's savings. Now, FPB can easily answer FA the range to which S belongs, without disclosing any data about the savings of C on any of the individual banks. Similarly, no information about C's savings will be disclosed to B_1, B_2,…,B_k. This trivial scheme can be further extended to blind FPB of the ID of C, therefore, the actual information that FPB will know would not be linkable to any individual.

6.3.3 The Future of Big Data Profiling Under the GDPR

Unquestionably, big data algorithmic profiling techniques are a huge step toward knowledge production and innovation. Hildebrandt [64] had long ago envisioned that "*advanced profiling technologies generate knowledge and since knowledge is power, profiling changes the power relationships between the profilers and the profiled*". According to the Ethics Advisory Group, digitally generated profiles based on very large quantities of data are powerful and increasingly unaccountable. Furthermore, as Pasquale notes, "*profiling is big business*" [106]. Certainly a successful one, given the latest revelations in the Cambridge Analytica case regarding the manipulation of 50 million Facebook profiles claimed to have won the 2016 US elections.[56] Yet, this fragile and complex balance between individual rights and collective knowledge should not be exclusively entrusted to market dynamics. Therefore, Pasquale concludes that the "*need to anticipate* [and regulate] *how profiling technologies categorize and preempt us is indeed more urgent than the need to prevent identification or to remain anonymous*" [106]. Solove [131] prophetically quoted a decade ago that "*protecting individuals from excessive observation, scrutiny, and categorization is not an individualistic agenda, but rather one of promoting societal goods*".

In that respect, the GDPR renders private and public sector more accountable to individuals and consequently challenges current industry and state approaches in terms of their privacy intrusive profiling practices. As a matter of fact, in 2018 the European Parliament, taking into account and citing the GDPR principles, published its motion for a resolution[57] on the Cambridge Analytica case in which emphasizes the need for much greater algorithmic accountability and transparency with regard to data processing and analytics by the private and public sectors. It also stresses that profiling based on online behaviour, socio-economic or demographic factors, for political and electoral purposes, should be prohibited.

Yet, the enforcement of the GDPR compelled many thus far established, or even recently enforced, international policies and legislations on tax and financial sector to be re-evaluated regarding their compatibility with its data protection principles [113]. In the tax domain, the GDPR's collision with the AEOI initiatives will certainly occupy extensively the interest of future law makers. In fact, the first legal complaint against the HMRC and the OECD CRS for infringing privacy and data

---

[56] https://www.theguardian.com/uk-news/2018/mar/23/leaked-cambridge-analyticas-blueprint-for-trump-victory.

[57] http://www.europarl.europa.eu/sides/getDoc.do?type=MOTION&reference=B8-2018-0480&format=XML&language=EN.

protection rights was filed in August 2018.[58], [59] Almost a month before that, the the European Parliament released a resolution,[60], [61] following a respective motion,[62] asking the European Commission to ensure that privacy and data protection rights are respected in the context of FATCA and the automatic exchange of tax data. The resolution asks, among others, the MSs to review their IGAs and to amend them, if necessary, in order to align them with the rights and principles of the GDPR. It also calls on the Commission to conduct a full assessment of the impact of FATCA and the US extraterritorial practice on EU citizens, EU financial institutions and EU economies, and regrets the inherent lack of reciprocity of IGAs signed by MSs, especially in terms of the scope of information to be exchanged, which is broader for MSs than it is for the US. Remarkably, the resolution calls on all MSs to collectively suspend the application of their IGAs until the US agrees to a multilateral approach to the AEOI, by either repealing FATCA and joining the CRS or renegotiating FATCA on an EU-wide basis and with identical reciprocal sharing obligations on both sides of the Atlantic. This resolution came as no surprise since two months earlier the Parliament's Policy Department for Citizens' Rights and Constitutional Affairs published a study[63], [64] on FATCA's compatibility with the EU legislation, and specifically the GDPR, and raised key points indicating FATCA's violation of the new EU legislation. The study pleaded, among others, for IGAs modification to align with the GDPR and to become truly reciprocal.

The future of PSD2/MiFID2 enforcement in the GDPR era is not reassuring either, as there is currently a lack of guidance on the implementation of both directives in order to be GDPR-compliant.[65] On top, being both directives, as opposed to the GDPR regulation status and its highly imposed fines, weakens any penalties that are to be determined in case of non-compliance with these initiatives. Hence, unless specific and detailed directions on their implementation are not timely provided as well as their coordination with the GDPR is not carefully regulated, their coexistence with the GDPR is uncertain.

---

[58] https://www.theguardian.com/money/2018/aug/02/mishcon-de-reya-complains-about-anti-tax-evasion-measures.

[59] https://globaldatareview.com/article/1172676/tax-and-money-laundering-information-schemes-face-gdpr-complaint.

[60] http://www.europarl.europa.eu/oeil/popups/ficheprocedure.do?lang=en&reference=2018/2646(RSP).

[61] http://www.europarl.europa.eu/news/en/press-room/20180628IPR06837/meps-want-to-open-negotiations-on-an-eu-us-fatca-agreement.

[62] http://www.europarl.europa.eu/sides/getDoc.do?type=MOTION&reference=B8-2018-0306&language=EN.

[63] http://www.europarl.europa.eu/RegData/etudes/STUD/2018/604967/IPOL_STU(2018)604967_EN.pdf.

[64] https://iapp.org/news/a/study-examines-fatca-through-the-lens-of-gdpr/.

[65] https://www.insideprivacy.com/financial-institutions/overlap-between-the-gdpr-and-psd2/.

Apart from the effect of the GDPR on tax and financial policies, its extraterritorial impact on the transfer of personal data outside the EU/EEA domain is also substantial. Currently, the Privacy Shield agreement, the framework for regulating transatlantic exchanges of personal data between the EU and US, is widely challenged[66] due to the US failure to protect personal data belonging to EU citizens. On 5 July 2018 the European Parliament, in the light of the Cambridge Analytica scandal, adopted a resolution[67] that stresses, among others, its concerns about the lack of specific rules and guarantees in the Privacy Shield for decisions based on automated processing and profiling, and called on the Commission to consider suspending its validity until the US authorities be fully compliant with the framework, setting a deadline of 1 September 2018 for this to be achieved. Following these developments, the status of the EU-US Privacy Shield became unclear.[68][69] However, according to European Commission's 2019 report on the third annual review of the functioning of the EU-US Privacy Shield, the framework works adequately but a number of concrete steps need to be taken to better ensure the effective functioning of the Privacy Shield in practice.[70]

Even though most EU policies aim to promote effective political and legal responses for enabling an innovative, transparent and with equal opportunities economic environment, most of the times they disregard data protection values and their impact to people's privacy, especially when these policies are combined with big data technology and algorithmic processing for profiling citizens and consumers. The widespread belief that public administrations, as they are held to a higher standard than the private sector organizations, can't engage in an identical implementation of big data profiling or automated-decision making systems[71] does not seem realistic anymore following the practices described in this Chapter. Although up to now the EU Data Protection Authorities haven't received any significant number of complaints on profiling, probably due to the novelty of the use of automated profiling and to a general lack of awareness by the citizenry [17], we firmly believe that this will not be the case henceforth.

---

[66] https://www.reuters.com/article/us-eu-dataprotection-usa/eu-u-s-personal-data-pact-faces-second-legal-challenge-from-privacy-groups-idUSKBN12X253?il=0.

[67] http://www.europarl.europa.eu/sides/getDoc.do?type=TA&reference=P8-TA-2018-0315&language=EN&ring=B8-2018-0305.

[68] https://www.euractiv.com/section/digital/news/eu-us-privacy-shield-review-jourova-to-meet-us-secretary-amid-compliance-concerns/.

[69] https://www.ictsd.org/bridges-news/bridges/news/eu-parliament-questions-us-compliance-with-data-privacy-shield-calls-for.

[70] https://ec.europa.eu/info/sites/info/files/report_on_the_third_annual_review_of_the_eu_us_privacy_shield_2019.pdf.

[71] https://bureaudehelling.nl/artikel-tijdschrift/efficiency-vs-accountability.

References

1. A. Acquisti, C. Taylor, L. Wagman, The economics of privacy. J. Econ. Lit. **54**(2), 442–492 (2016)
2. M. Ananny, K. Crawford, Seeing without knowing: limitations of the transparency ideal and its application to algorithmic accountability. New Media Soc. **20**(3):973–989 (2018)
3. Article 29 Data Protection Working Party, Letter 21/06/2012 to the Director General of Taxation and Customs Union European Commission Ref. Ares (2012) 746461 following a request for assistance by DG TAXUD to evaluate the compatibility of the obligations under US Foreign Account Tax Compliance Act (FATCA) and Directive 95/46/EC (2012). https://ec.europa.eu/justice/article-29/documentation/other-document/files/2012/20120621_letter_to_taxud_fatca_en.pdf
4. Article 29 Data Protection Working Party, Guidelines on the right to data portability, WP242rev.01, Adopted on 13 December 2016 (2017). As last Revised and adopted on 5 April 2017. https://ec.europa.eu/newsroom/document.cfm?doc_id=44099
5. Article 29 Data Protection Working Party, Guidelines on Automated individual decision-making and Profiling for the purposes of Regulation 2016/679, WP251rev.01, Adopted on 3 October 2017. As last Revised and Adopted on 6 February 2018 (2018). http://ec.europa.eu/newsroom/article29/item-detail.cfm?item_id=612053
6. Y. Baimbetov, I, Khalil I, M. Steinbauer, G. Anderst-Kotsis, Using big data for emotionally intelligent mobile services through multi-modal emotion recognition, /textitInternational Conference on Smart Homes and Health Telematics (Springer, Berlin), pp. 127–138 (2015)
7. P. Baker, CRS/DAC, FATCA and the GDPR. Br. Tax Rev. **3**, 249–252 (2016)
8. P. Baker, P. Pistone, BEPS Action 16: the taxpayers' right to an effective legal remedy under European law in cross-border situations. EC Tax Rev. **25**(5), 335–345 (2016)
9. D.E. Bakken, R. Rarameswaran, D.M. Blough, A.A. Franz, T.J. Palmer, Data obfuscation: anonymity and desensitization of usable data sets. IEEE Secur. Priv. **2**(6), 34–41 (2004)
10. S. Barocas, H. Nissenbaum, Big data's end run around procedural privacy protections. Commun. ACM **57**(11), 31–33 (2014)
11. S. Barocas, A.D. Selbst, Big data's disparate impact. Calif. L. Rev. **104**, 671 (2016)
12. M. Bateson, D. Nettle, G. Roberts, Cues of being watched enhance cooperation in a real-world setting. Biol. Lett. **2**(3), 412–414 (2006)
13. P. Bessard, Inidividual rights and tax oppression in the OECD. Liberales Institut paper **3**, 1–29 (2017)
14. C. Bettini, D. Riboni, Privacy protection in pervasive systems: State of the art and technical challenges. Pervasive Mob. Comput. **17**, 159–174 (2015)
15. D. Boneh, B. Waters, Conjunctive, subset, and range queries on encrypted data, in/textitTheory of Cryptography Conference (Springer, Berlin, 2007), pp. 535–554
16. J.W. Bos, K. Lauter, M. Naehrig, Private predictive analysis on encrypted medical data. J. Biomed. Inf. **50**, 234–243 (2014)
17. F. Bosco, N. Creemers, V. Ferraris, D. Guagnin, B.J. Koops, Profiling technologies and fundamental rights and values: regulatory challenges and perspectives from European data protection authorities. In: *Reforming European Data Protection Law* (Springer, Berlin), pp. 3–33
18. R. Bost, R.A. Popa, S. Tu, S. Goldwasser, Machine learning classification over encrypted data. In: NDSS, vol 4324 (2015), p. 4325
19. J. Brickell, V. Shmatikov, Privacy-preserving classifier learning, in /textitInternational Conference on Financial Cryptography and Data Security (Springer, Berlin, 2009), pp. 128–147
20. D. Brin, *The Transparent Society: Will Technology Force Us to Choose Between Privacy and Freedom?* (Basic Books, 1999)
21. J. Burrell, How the machine "thinks": Understanding opacity in machine learning algorithms. Big Data Soc. **3**(1), 2053951715622512 (2016)
22. L.A. Bygrave, Automated profiling: minding the machine: Article 15 of the EC data protection directive and automated profiling. Comput. Law Secu. Rev. **17**(1), 17–24 (2001)

23. C.S. Calude, G. Longo, The deluge of spurious correlations in big data. Found. Sci. **22**(3), 595–612 (2017)
24. A. Campbell, T. Choudhury, From smart to cognitive phones. IEEE Pervasive Comput. **3**(11), 7–11 (2012)
25. J. Chen, A. Bauman, M. Allman-Farinelli, A study to determine the most popular lifestyle smartphone applications and willingness of the public to share their personal data for health research. Telemed. e-Health **22**(8), 655–665 (2016)
26. D.K. Citron, Technological due process. Wash Univ. Law Rev. **85**, 1249 (2007)
27. D.K. Citron, F. Pasquale, The scored society: due process for automated predictions. Wash Law Rev. **89**, 1 (2014)
28. A.J. Cockfield, Protecting taxpayer privacy rights under enhanced cross-border tax information exchange: toward a multilateral taxpayer bill of rights. UBC Law Rev. **42**(2), 421 (2010)
29. A.J. Cockfield, Bid data and tax haven secrecy. Fla Tax Rev. **18**, 483 (2015)
30. J.E. Cohen, What privacy is for. Harv Law Rev. **126**, 1904 (2012)
31. F.T. Commission et al., Big data: a tool for inclusion or exclusion? Understanding the issues. FTC Report, (2016, January)
32. Council of Europe (2010) The protection of individuals with regard to automatic processing of personal data in the context of profiling. Recommendation CM/Rec(2010)13 and explanatory memorandum, 23 November 2010. https://rm.coe.int/16807096c3
33. Council of Europe (2017) Guidelines on the protection of individuals with regard to the processing of personal data in a world of Big Data". https://rm.coe.int/16806ebe7a
34. R. Cowie, Ethical issues in affective computing. The Oxford Handbook of Affective Computing (2015), p. 334
35. K. Crawford, J. Schultz, Big data and due process: toward a framework to redress predictive privacy harms. BCL Rev. **55**, 93 (2014)
36. G. D'Acquisto, J. Domingo-Ferrer, P. Kikiras, V. Torra, Y.A. de Montjoye, A. Bourka, Privacy by design in big data: an overview of privacy enhancing technologies in the era of big data analytics. Eur. Union Agency Netw. Inf. Secur. Retrieved from https://www.enisa.europa.eu/publications/big-data-protection, version 1.0 (2015)
37. J.P. Daries, J. Reich, J. Waldo, E.M. Young, J. Whittinghill, A.D. Ho, D.T. Seaton, I. Chuang, Privacy, anonymity, and big data in the social sciences. Commun. ACM **57**(9), 56–63 (2014)
38. L. De Simone, R. Lester, K. Markle, Transparency and tax evasion: evidence from the foreign account tax compliance act (FATCA). J. Account. Res. **58**(1), 105–153 (2020)
39. F. Debelva, I. Mosquera, Privacy and confidentiality in exchange of information procedures: some uncertainties, many issues, but few solutions. Intertax **45**(5), 362–381 (2017)
40. N. Diakopoulos, Accountability in algorithmic decision making. Commun. ACM **59**(2), 56–62 (2016)
41. N. Diepvens, F. Debelva, The evolution of the exchange of information in direct tax matters: the taxpayer's rights under pressure. EC Tax Rev. **24**(4), 210–219 (2015)
42. F. Doshi-Velez, M. Kortz, R. Budish, C. Bavitz, S. Gershman, D. O'Brien, S. Schieber, J. Waldo, D. Weinberger, A. Wood, Accountability of AI under the law: The role of explanation (2017). arXiv preprint arXiv:171101134
43. C. Dwork, Differential privacy, in Automata, Languages and Programming. ICALP 2006. Lecture Notes in Computer Science, vol. 4052 (Springer, Berlin, Heidelberg, 2006), pp. 1–12
44. C. Dwork, M. Hardt, T. Pitassi, O. Reingold, R. Zemel, Fairness through awareness, in *Proceedings of the 3rd Innovations in Theoretical Computer Science Conference* (ACM, 2012), pp. 214–226
45. L. Edwards, Privacy, security and data protection in smart cities: a critical EU law perspective. Eur. Data Prot. Law Rev. **2**, 28 (2016)
46. L. Edwards, M. Veale, Slave to the algorithm: why a right to an explanation is probably not the remedy you are looking for. Duke Law Tech. Rev. **16**, 18 (2017)
47. M. Egele, C. Kruegel, E. Kirda, G. Vigna, PiOS: Detecting privacy leaks in iOS applications, in *NDSS,2011*, pp. 177–183

48. L. Einav, J. Levin, The data revolution and economic analysis. Innov. Policy Econ. **14**(1), 1–24 (2014)
49. Z. Erkin, J.R. Troncoso-Pastoriza, R.L. Lagendijk, F. Pérez-González, Privacy-preserving data aggregation in smart metering systems: an overview. IEEE Signal Process. Mag. **30**(2), 75–86 (2013)
50. S. Gadžo, I. Klemenčić, Effective international information exchange as a key element of modern tax systems: promises and pitfalls of the OECD's common reporting standard. Publ. Sect. Econ. **41**(2), 207–226 (2017)
51. R. Gilad-Bachrach, N. Dowlin, K. Laine, K. Lauter, M. Naehrig, J. Wernsing, Cryptonets: applying neural networks to encrypted data with high throughput and accuracy, in *International Conference on Machine Learning, 2016*, pp. 201–210
52. A. Goldfarb, C. Tucker, Privacy and innovation. Innov. Policy Econ. **12**(1), 65–90 (2012)
53. B. Goodman, S. Flaxman, European union regulations on algorithmic decision-making and a "right to explanation." AI Mag. **38**(3), 50–57 (2017)
54. S.D. Gosling, W. Mason, Internet research in psychology. Ann. Rev. Psychol. **66**, 877–902 (2015)
55. T. Graepel, K. Lauter, M. Naehrig, Ml confidential: Machine learning on encrypted data, in *International Conference on Information Security and Cryptology* (Springer, Berlin, 2012), pp. 1–21
56. V. Griskevicius, J.M. Tybur, B. Van den Bergh, Going green to be seen: status, reputation, and conspicuous conservation. J. Pers. Soc. Psychol **98**(3), 392 (2010)
57. S. Gutwirth, M. Hildebrandt, Some caveats on profiling, in *Data Protection in a Profiled World* (Springer, Berlin), pp. 31–41
58. M. Gymrek, A.L. McGuire, D. Golan, E. Halperin, Y. Erlich, Identifying personal genomes by surname inference. Science **339**(6117), 321–324 (2013)
59. M. Hatfield, Taxation and surveillance: an agenda. Yale J. Law Technol. **2014**, 34 (2015)
60. M. Hatfield, Privacy in taxation. Florida State Univ. Law Rev. **44**, 579 (2016)
61. J. Hemerly, Public policy considerations for data-driven innovation. Computer **46**(6), 25–31 (2013)
62. E. Hesamifard, H. Takabi, M. Ghasemi, R.N. Wright, Privacy-preserving machine learning as a service. Proc. Priv. Enhancing Technol. **3**, 123–142 (2018)
63. M. Hildebrandt, Defining profiling: a new type of knowledge? in *Profiling the European Citizen* (Springer, Berlin, 2008a), pp. 17–45
64. M. Hildebrandt, Profiling and the rule of law. Identity Inf. Soc. **1**(1), 55–70 (2008)
65. M. Hildebrandt, Who is profiling who? Invisible visibility, in *Reinventing Data Protection?* (Springer, Berlin, 2009), pp. 239–252
66. M. Hildebrandt, Smart Technologies and the End (s) of Law: Novel Entanglements of Law and Technology (Edward Elgar Publishing, 2015)
67. M. Hildebrandt, Privacy as protection of the incomputable self: From agnostic to agonistic machine learning. Theor. Inquiries Law **20**(1), 83–121 (2019)
68. C. HJI Panayi, Current trends on automatic exchange of information. Singapore Management University School of Accountancy Research Paper **(2016-S)**, 43 (2016)
69. E. Horvitz, D. Mulligan, Data, privacy, and the greater good. Science **349**(6245), 253–255 (2015)
70. K.A. Houser, D. Sanders, The use of big data analytics by the IRS: efficient solutions or the end of privacy as we know it. Vand J. Enter. Tech. Law **19**, 817 (2016)
71. M. Hurley, J. Adebayo, Credit scoring in the era of big data. Yale J. Law Tech. **18**, 148 (2016)
72. F. Kaltheuner, E. Bietti, Data is power: Towards additional guidance on profiling and automated decision-making in the GDPR. J. Inf. Rights Policy Pract. **2**(2) (2018)
73. D. Kamarinou, C. Millard, J. Singh, Machine learning with personal data. Queen Mary School of Law Legal Studies Research Paper (247) (2016)
74. M.E. Kaminski, The right to explanation, explained. University of Colorado Law Legal Studies Research Paper No 18-24; Berkeley Technology Law J. **34** (2018)

75. A. Kapadia, D. Kotz, N. Triandopoulos, Opportunistic sensing: security challenges for the new paradigm, in *first International Communication Systems and Networks and Workshops* (IEEE, 2009), pp. 1–10
76. M. Kosinski, D. Stillwell, T. Graepel, Private traits and attributes are predictable from digital records of human behavior. Proc. Natl. Acad. Sci. **110**(15), 5802–5805 (2013)
77. J.A. Kroll, S. Barocas, E.W. Felten, J.R. Reidenberg, D.G. Robinson, H. Yu, Accountable algorithms. U Pa Law Rev. **165**, 633 (2016)
78. K. Kursawe, G. Danezis, M. Kohlweiss, Privacy-friendly aggregation for the smart-grid, in *International Symposium on Privacy Enhancing Technologies Symposium* (Springer, Berlin, 2016), pp. 175–191
79. H. Lakkaraju, S.H. Bach, J. Leskovec, Interpretable decision sets: a joint framework for description and prediction, in *Proceedings of the 22nd ACM SIGKDD International Conference on Knowledge Discovery and Data Mining* (ACM, 2016), pp. 1675–1684
80. B. Letham, C. Rudin, T.H. McCormick, D. Madigan et al., Interpretable classifiers using rules and Bayesian analysis: Building a better stroke prediction model. Ann. Appl. Stat. **9**(3), 1350–1371 (2015)
81. N. Li, T. Li, S. Venkatasubramanian, t-closeness: Privacy beyond k-anonymity and l-diversity, in *2007 IEEE 23rd International Conference on Data Engineering* (IEEE, 2007), pp 106–115
82. P. Li, T. Li, H. Ye, J. Li, X. Chen, Y. Xiang, Privacy-preserving machine learning with multiple data providers. Future Gener. Comput. Syst. **87**, 341–350 (2018)
83. Q. Li, G. Cao, T.F. La Porta, Efficient and privacy-aware data aggregation in mobile sensing. IEEE Trans. Dependable Secure Comput. **11**(2), 115–129 (2014)
84. R. Li, A.X. Liu, A.L. Wang, B. Bruhadeshwar, Fast range query processing with strong privacy protection for cloud computing. Proc. VLDB Endowment **7**(14), 1953–1964 (2014)
85. Z.C. Lipton, The Mythos of Model Interpretability. Queue **16**(3), 30:31–30:57 (2018)
86. W. Lu, J. Sakuma,More practical privacy-preserving machine learning as a service via efficient secure matrix multiplication, in *Proceedings of the 6th Workshop on Encrypted Computing & Applied Homomorphic Cryptography* (ACM, 2018), pp. 25–36
87. A. Machanavajjhala, D. Kifer, J. Gehrke, M. Venkitasubramaniam, l-diversity: privacy beyond k-anonymity. ACM Trans. Knowl. Disc. Data (TKDD) **1**(1), 3 (2007)
88. H. Malekian, Profiling under General Data Protection Regulation (GDPR): Stricter Regime? (2016). https://www.linkedin.com/pulse/profiling-under-general-data-protection-regulation-gdpr-malekian
89. A. Mantelero, Personal data for decisional purposes in the age of analytics: from an individual to a collective dimension of data protection. Comput. Law Secur. Rev. **32**(2), 238–255 (2016)
90. A. Mantelero, Regulating big data. The guidelines of the Council of Europe in the context of the European data protection framework. Comput. Law Secur. Revi. **33**(5), 584–602 (2017)
91. V. Mayer-Schönberger, K. Cukier, *Big Data: A Revolution that Will Transform How We Live, Work, and Think* (Houghton Mifflin Harcourt, 2013)
92. V. Mayer-Shönberger, Y. Padova, Regime change: enabling big data through Europe's new data protection regulation. Colum Sci. Tech Law Rev. **17**, 315 (2015)
93. D. McMillan, A. Morrison, M. Chalmers, Categorised ethical guidelines for large scale mobile HCI, in *Proceedings of the SIGCHI Conference on Human Factors in Computing Systems* (ACM, 2013), pp. 1853–1862
94. I. Mendoza, L.A. Bygrave, The right not to be subject to automated decisions based on profiling, in *EU Internet Law* (Springer, Berlin, 2017), pp. 77–98
95. G. Miller, The smartphone psychology manifesto. Perspectives Psychol. Sci. **7**(3), 221–237 (2012)
96. B.D. Mittelstadt, L. Floridi, The ethics of big data: current and foreseeable issues in biomedical contexts. Sci. Eng. Ethics **22**(2), 303–341 (2016)
97. P. Mohassel, Y. Zhang, Secureml: a system for scalable privacy-preserving machine learning, in *2017 IEEE Symposium on Security and Privacy (SP)* (IEEE, 2017), pp. 19–38
98. Narayanan A, Shmatikov V (2008) Robust de-anonymization of large sparse datasets, in IEEE Symposium on Security and Privacy, 2008. SP 2008, IEEE, pp 111–125

99. A. Narayanan, V. Shmatikov, Myths and fallacies of personally identifiable information. Commun. ACM **53**(6), 24–26 (2010)
100. K. O'Hara, N. Shadbolt, *The Spy in the Coffee Machine: The End of Privacy as We Know It.* (Oneworld Publications, 2014)
101. P. Ohm, Broken promises of privacy: responding to the surprising failure of anonymization. UCLA law Rev. **57**, 1701 (2009)
102. P. Ohm, The fourth amendment in a world without privacy. Miss. Law J. **81**, 1309–1355 (2012)
103. O. Ohrimenko, F. Schuster, C. Fournet, A. Mehta, S. Nowozin, K. Vaswani, M. Costa, Oblivious multi-party machine learning on trusted processors, in *25th (|USENIX|) Security Symposium (|USENIX|) Security 16* (2016), pp. 619–636
104. M. Oostveen, K. Irion, The golden age of personal data: How to regulate an enabling fundamental right?, in *Personal Data in Competition*. (Springer, Consumer Protection and Intellectual Property Law, 2018), pp. 7–26
105. A. Papageorgiou, M. Strigkos, E. Politou, E. Alepis, A. Solanas, C. Patsakis, Security and privacy analysis of mobile health applications: the alarming state of practice. IEEE Access **6**, 9390–9403 (2018)
106. F. Pasquale, *The Black Box Society* (Harvard University Press, 2015)
107. C. Patsakis, P. Laird, M. Clear, M. Bouroche, A. Solanas, Interoperable privacy-aware e-participation within smart cities. Computer **48**(1), 52–58 (2015)
108. V. Pejovic, M. Musolesi, Anticipatory mobile computing for behaviour change interventions, in *Proceedings of the 2014 ACM International Joint Conference on Pervasive and Ubiquitous Computing: Adjunct Publication* (ACM, 2014), pp. 1025–1034
109. V. Pejovic, M. Musolesi, Anticipatory mobile computing: a survey of the state of the art and research challenges. ACM Comput. Surv. (CSUR) **47**(3), 47 (2015)
110. R.W. Picard, Affective computing: from laughter to IEEE. IEEE Trans. Affect. Comput. **1**(1), 11–17 (2010)
111. E. Politou, E. Alepis, C. Patsakis, A survey on mobile affective computing. Comput. Sci. Rev. **25**, 79–100 (2017)
112. E. Politou, E. Alepis, C. Patsakis, Profiling tax and financial behaviour with big data under the GDPR. Comput. Law Secur. Rev. **35**(3), 306–329 (2019)
113. E. Politou, E. Alepis, C. Patsakis, Profiling tax and financial behaviour with big data under the GDPR. Comput. Law Secur. Rev. **35**(3), 306–329 (2019)
114. M. Raento, A. Oulasvirta, N. Eagle, Smartphones an emerging tool for social scientists. Sociolo. Methods Res. **37**(3), 426–454 (2009)
115. J. Reijers, B. Jacobs, I.E. Poll, Payment Service Directive 2. Ph.D. thesis, Thesis for the Degree of Master of Science in Information Sciences at the Radboud University Nijmegen, The Netherlands (2016)
116. M.T. Ribeiro, S. Singh, C. Guestri, Why should i trust you? Explaining the predictions of any classifier, in *Proceedings of the 22nd ACM SIGKDD International Conference on Knowledge Discovery and Data Mining* (ACM, 2016), pp. 1135–1144
117. N.M. Richards, J.H. King, Three paradoxes of big data. Stanford Law Rev. Online **66**, 41 (2013)
118. S.A. Rocha, Exchange of Tax-Related Information and the Protection of Taxpayer Rights: General Comments and the Brazilian Perspective. Bull. Int. Taxation (2016), pp. 502–16
119. A. Rouvroy , "Of data and men". Fundamental rights and freedoms in a world of big data. In: Bureau of the Consultative Committee of the Convention for the Protection of Individuals with Regard to Automatic Processing of Personal Data [ETS 108] (2016)
120. I.S. Rubinstein, Big data: the end of privacy or a new beginning? Int. Data Priv. Law **3**(2), 74–87 (2013)
121. S. Samet, A. Miri, Privacy-preserving back-propagation and extreme learning machine algorithms. Data Knowl. Eng. **79**, 40–61 (2012)
122. A. Savin, Profiling and automated decision making in the present and new EU data protection frameworks (2013) **14**, 1. http://openarchivecbsdk/bitstream/handle/10398/89

123. M. Schaper, Data protection rights and tax information Exchange in the European Union: an uneasy combination. Maastricht J. Eur. Comp. Law **23**(3), 514–530 (2016)
124. B.W. Schermer, The limits of privacy in automated profiling and data mining. Comput. Law Sec. Rev. **27**(1), 45–52 (2011)
125. P.M. Schwartz, D.J. Solove, Reconciling personal information in the United States and European Union. Calif Law Rev. **102**, 877 (2014)
126. A.D. Selbst, J. Powles, Meaningful information and the right to explanation. Int. Data Priv. Law **7**(4), 233–242 (2017)
127. J.C. Sharman, Privacy as roguery: personal financial information in an age of transparency. Pub. Adm. **87**(4), 717–731 (2009)
128. D. Shenk, *Data Smog: Surviving the Information Glut* (Harper San Francisco) (1998)
129. R. Shokri, V. Shmatikov, Privacy-preserving deep learning, in *Proceedings of the 22nd ACM SIGSAC Conference on Computer and Communications Security*, ACM (2015), pp. 1310–1321
130. D.J. Solove, A taxonomy of privacy. University of Pennsylvania Law Rev. **154**, 477–564 (2006)
131. D.J. Solove, I've got nothing to hide and other misunderstandings of privacy. San Diego Law Rev. **44**, 745 (2007)
132. M. Somare, V. Wöhrer, Automatic exchange of financial information under the directive on administrative coopération in the light of the global movement towards transparency. Intertax **43**(12), 804–815 (2015)
133. C. Spensky, J. Stewart, A. Yerukhimovich, R. Shay, A. Trachtenberg, R. Housley, R.K. Cunningham, SoK: privacy on mobile devices—it's complicated. Proc. Priv. Enhancing Technol. **3**, 96–116 (2016)
134. S. Spiekermann, A. Acquisti, R. Böhme, K.L. Hui, The challenges of personal data markets and privacy. Electron. Markets **25**(2), 161–167 (2015)
135. J. Staiano, N. Oliver, B. Lepri, R. de Oliveira, M. Caraviello, N. Sebe, Money walks: a human-centric study on the economics of personal mobile data, in *Proceedings of the 2014 ACM International Joint Conference on Pervasive and Ubiquitous Computing* (ACM, 2014), pp. 583–594
136. L. Sweeney, Simple demographics often identify people uniquely. Health (San Francisco) **671**, 1–34 (2000)
137. L. Sweeney, k-anonymity: a model for protecting privacy. Int. J. Uncertainty Fuzziness Knowl. Based Syst. **10**(05), 557–570 (2002)
138. L. Sweeney, Matching known patients to health records in Washington State data (2013). arXiv preprint arXiv:13071370
139. L. Taylor, R. Schroeder, E. Meyer, Emerging practices and perspectives on big data analysis in economics: bigger and better or more of the same? Big Data Soc. **1**(2), 2053951714536877 (2014)
140. L. Taylor, L. Floridi, B. Van der Sloot, *Group Privacy: New Challenges of Data Technologies*, vol. 126 (Springer, Berlin, 2016)
141. C.P. Tello, FATCA: catalyst for global cooperation on exchange of tax information. Bull. Int. Taxation 68 (2014)
142. O. Tene, J. Polonetsky, Big data for all: privacy and user control in the age of analytics. Nw J. Tech. & Intell. Prop. **11**, xxvii (2012)
143. A.B. Thimmesch, Tax privacy. Temp Law Rev. **90**, 375 (2017)
144. A. Tutt, An FDA for algorithms. Admin Law Rev. **83** (2017)
145. UK Information Commissionerś Office (UK ICO), Big data, artificial intelligence, machine learning and data protection (2017). https://ico.org.uk/media/for-organisations/documents/2013559/big-data-ai-ml-and-data-protection.pdf
146. L. Urquhart, N. Sailaja, D. McAuley, Realising the right to data portability for the domestic internet of things. Personal Ubiquitous Compu. **22**(2), 317–332 (2018)

147. B. Van Alsenoy, V. Verdoodt, R. Heyman, E. Wauters, J. Ausloos, G. Acar, From social media service to advertising network: a critical analysis of Facebooks revised policies and terms (v1.3)—Report for the Belgian Privacy Commission on Facebook's revised Data Use Policy. Brussels, Belgium, Retrieved from http://www.law.kuleuven.be/icri/en/news/item/facebooks-revised-policies-and-terms-v1-3.pdf (2015)
148. M. Veale, M. Van Kleek, R. Binns, Fairness and accountability design needs for algorithmic support in high-stakes public sector decision-making, in *Proceedings of the 2018 chi Conference on Human Factors in Computing Systems* (ACM, 2018), p. 440
149. S. Wachter, B. Mittelstadt, L. Floridi, Why a right to explanation of automated decision-making does not exist in the general data protection regulation. Int. Data Priv. Law **7**(2), 76–99 (2017)
150. S. Wachter, B. Mittelstadt, C. Russell, Counterfactual explanations without opening the black box: automated decisions and the GPDR. Harv J. Law Tech. **31**, 841 (2017)
151. S. Yu, Big privacy: challenges and opportunities of privacy study in the age of big data. IEEE Access **4**, 2751–2763 (2016)
152. T.Z. Zarsky, Transparent predictions. Univ. Ill Law Rev. 1503 (2013)
153. T.Z. Zarsky, Incompatible: the GDPR in the age of big data. Seton Hall Lae Rev. **47**, 995 (2016)

Chapter 7
Privacy in Blockchain

Abstract Privacy in blockchains is rather complicated as it contradicts with some highly praised properties of blockchain such as immutability. Immutability is considered a cornerstone of blockchains' security and, therefore, an indisputable property according to which transactional blockchain data cannot be edited nor deleted. However, blockchain's immutability is being called into question lately in the light of the new erasing requirements imposed by the GDPR's "Right to be Forgotten (RtbF)" provision. Given that the RtbF compels data stored in blockchains to be editable so as restricted content redactions, modifications or deletions to be carried out when requested, blockchains' alignment with the regulation is indeed challenging, if not unfeasible. Towards resolving this discrepancy, in this Chapter we first discuss the privacy challenges faced by blockchain technology, and we then explore blockchain's contradiction with the RtbF erasing provisions of the GDPR. In this respect, we provide a comprehensive review on the state-of-the-art approaches, technical methods and workarounds and advanced cryptographic techniques that have been put forward to resolve this contradiction, and we discuss their potentials and limitations when applied at large to either permissioned or permissionless blockchains.

7.1 Introduction

As discussed in Sect. 5.3.1.3, a highly praised property that underpins blockchain's secure and transparent nature, is immutability. Blockchain's immutability, according to which tampering with blockchain data is practically impossible, guarantees its transactional integrity, security and censorship-resistance nature. Nevertheless, this property, albeit desirable in some contexts, contradicts several privacy requirements and data protection rights when personal data are at stake. Among others, it obviously challenges the "*Right to be Forgotten*" (RtbF) enshrined in the GDPR according to which individuals have the right to delete their personal data if certain conditions apply [1, 2]. As the RtbF obliges blockchain data to be editable in order restricted content redactions, modifications or deletions to be applied when requested, blockchains compliance with the regulation is indeed challenging, if not impracticable.

E. Politou et al., *Privacy and Data Protection Challenges in the Distributed Era*,
Learning and Analytics in Intelligent Systems 26,
https://doi.org/10.1007/978-3-030-85443-0_7

In this Chapter, we discuss the privacy challenges faced by blockchain technology, and we explore its contradiction with the RtbF erasing provisions of the GDPR. Towards resolving this conflict, various methods and techniques for mutable blockchains have been proposed to satisfy regulatory forgetting requirements while preserving blockchains' security and integrity. To this end, we provide a comprehensive review on these technical workarounds and advanced cryptographic techniques and we discuss their potentials, constraints and limitations when applied in the wild to either permissioned or permissionless blockchains.

7.2 Blockchain Privacy

By design, blockchains are based on the principle of complete transparency according to which transactions, even if they are hashed or encrypted, are visible to all participating nodes so that they can be validated [3]. Therefore, since the content of every transaction is exposed to every node on the network, transactional privacy in blockchains is hard to be attained. Nevertheless, in permissioned blockchains, where the nodes are known, privacy and confidentiality are usually preserved much more efficiently than in permissionless settings through the use of access control policies. On the other hand, while user accounts in permissionless blockchains can largely stay anonymous, and as such are thought to provide a series of privacy benefits to their users, many studies have demonstrated that there are still considerable risks to users' privacy [4–8]. For instance, research has shown that even when users are hiding behind multiple pseudonyms, these can be correlated and often identify them [7, 9–11]. Adding to this the fact that transactions are linked, one can retrieve the full history of all transactions performed on a blockchain [11].

Due to the transparent and permanent nature of blockchain technology which requires data to be stored forever and to be publicly available to the entire network, putting personal data on blockchains has been broadly discouraged. As it has been argued, storing personal data into blockchains it is like having again "Cambridge Analytica"—a severe surveillance scandal—but on the blockchain [12]. However, blockchains do not have to expose personal data directly to reveal individuals' personal information. By exploiting metadata information and by applying big data analytics, potentially sensitive information can also be retrieved, e.g. recording visits to health practitioners may reveal sensitive details on someone's health status [8]. As it has been demonstrated in the literature, achieving privacy in a lightweight and flexible manner for all DLTs, in general, is still an open research question [6]. That being said, it is worth noting that privacy was never one of blockchain's original problems to be addressed. As Buterin, the founder of the ethereum blockchain, puts it "*blockchains do not solve privacy issues and are an authenticity solution only*" [13].

Despite this limitation, several approaches based on cryptographic techniques such as homomorphic encryption, zero-knowledge proofs [14], and secure Multi-Party Computation (MPC) [15] have been proposed to address transactional privacy in blockchains. Broadly speaking, these techniques enable specific computations to be performed without revealing the inputs and outputs of those computations. These methods, however, are resource-intensive, so it is almost impossible to be implemented at scale [3]. Tumblers or mixing services have also been used intensively lately as a means to provide strong notions of anonymity in public blockchain networks [16].

7.3 Blockchain's Immutability and the "Right to Be Forgotten"

Blockchains by definition are unable to forget since tampering with transactional data stored in blockchains has been identified as nearly impossible [17]. Undeniably, this immutable and transparent record-keeping of blockchain data facilitates the movement and storage of information in a secure, auditable and credible way, and consequently guarantees blockchains' credibility, persistence and security. Despite its apparent benefits, blockchain's immutability has also some unintended consequences such as when erroneous or illegal content is stored in the blockchain [18]. Likewise, as already discussed in Sect. 7.2, blockchain immutability presents several risks to people's privacy. More precisely, immutability's collision with privacy and data protection rights renders absolute immutability a major barrier to blockchain's adoption when personal data are at stake [19]. In this regard, immutability, a hitherto indisputable and highly advertised property and the cornerstone of blockchain's security, is being called into question in the light of the privacy requirements imposed by the recently adopted European data protection regulation, the GDPR. Although the GDPR provides strict requirements for the processing of the personal data and offers extended legal rights to individuals residing in the EU, in its provided recitals and articles it does not take into account decentralized technologies such as DLTs and blockchains. On the one hand, this is because regulators deliberately chose to follow a technology-agnostic approach in order not to bind he provisions of the law with current trends and state-of-the-art technologies in computer science [2]. On the other hand, however, this stems from the fact that over the long period under which the final GDPR text was being debated and finalized, blockchain technology has not been a widespread technological trend that it is nowadays. As a result, various legal and technical divergences and incompatibilities between the GDPR and the blockchain technology have been unavoidably identified [17, 19–22].

As discussed in 3.5.3, the most profound and controversial GDPR provision is the Article 17 that anticipates the RtbF which offers the possibility of individuals to request the erasure of their personal data when certain conditions are met (Article 17(1)). In particular, the RtbF entails the permanent erasure of personal data upon request and from all the places to which they have been disseminated [2]. Beyond any doubt, and as it has been already thoroughly analysed, the impact of encompassing the RtbF on contemporary information systems is immense, whereas its integration into the design of future technological developments is currently disputable [2, 23].

Blockchain technology, due to its immutability, is one such advanced development that contradicts the RtbF. Although one might argue that the anonymization of personal data residing in blockchains through public key cryptography is a reasonable step for blockchain data to fall outside of the scope of GDPR, it should be stressed that private and public keys as well as hashed data are pseudonymous, not anonymous, and therefore also qualify as personal data under the GDPR (Article 4(1)). Hence, as discussed in 6.3, pseudonymous data need to be protected as well [2, 17, 24]. In other words, blockchain compliance with the GDPR only through the use of hash values and public key cryptography cannot be guaranteed [19]. Taking further into account that data stored in blockchains are never completely anonymous (see 7.2), it is apparent that the RtbF strikes at the heart of the blockchain's immutability property.

Against this background, CNIL, the French Data Protection Authority, notes that it is technically impossible to grant the data subject's the right to request for erasure when data are entered in the blockchain. In fact, while the CNIL recognizes that there are some cryptographic methods that may result to the data being "almost inaccessible", it also questions the extent to which these solutions provide full compliance with the GDPR since they do not "*strictly speaking, result in an erasure of the data insofar as the data would still exist in the blockchain*" [25, 26]. Along the same lines, the European Data Protection Supervisor (EDPS) stresses the importance of enabling the manageability of the personal data, i.e. their alteration, deletion, and selective disclosure, as a mean to maintain people's privacy [27]. Furthermore, a recent resolution from the European Parliament on DLTs and blockchains raises the need for blockchain applications to be compliant with the GDPR, stressing the fact that the RtbF is not easily applicable to this technology [28].

Inevitably, the RtbF has been seen by many blockchain advocates and crypto activists as an obstacle for expanding the blockchain technology to a broad area of applications. Still, others have argued that approaches for adding preapproved, limited, and transparent methods to alter data on an immutable system is a trade-off necessary to be able to utilize the advantages of the blockchain technology [19, 21]. In this respect, the World Economic Forum has sounded the alarm about the struggling of blockchain innovation due to the GDPR and urged for flexible policy frameworks to allow the benefits of data and technology to be realised [29]. Ideally, for enabling data deletion, the participants of a blockchain would have to agree on an effective process to jointly execute a lawful request to erase personal data from the decentralized ledgers [30]. As already discussed, in permissioned blockchains where there are specific entities (authorities or enterprises) in charge and legally accountable, introducing mutability in the blockchain without interrupting its functionality should not be considered an impossible task [6, 31]. In this perspective and in the context of private permissioned blockchains, the term "pragmatic immutability" has been coined to pave the way for greater blockchain adoption outside the world of cryptocurrency [32].

However, introducing mutability in public permissionless blockchains is rather challenging due to the absolute lack of trust among the participants. Yet, there exist optimistic voices that put their faith in advanced cryptographic techniques to guarantee individual privacy in decentralized architectures such as blockchains [8]. With this in mind, several research works have been carried out lately in an attempt to conform blockchains to the RtbF erasing requirements and to consequently adjust them to privacy-intensive applications. Among others, these works include technical workarounds and advanced cryptographic methods to either bypass or remove blockchain immutability both for permissioned and permissionless blockchains. The state-of-the-art of these works is discussed hereafter.

7.4 Current Efforts for Balancing Immutability and the RtbF

To address privacy issues arising from blockchains, and particularly to tackle the controversy around blockchain's immutability and the RtbF, various approaches have been embraced by researchers and information scientists. These comprise technical methods broadly used nowadays to bypass the blockchain's conflict with the RtbF, as well as cryptographic and other advanced methods aiming at removing blockchain's immutability in order to conditional edit its data. An overview of these solutions is illustrated in Fig. 7.1.

In what follows we attempt, on the one hand, to summarize all these innovative methods and state-of-the-art techniques and, on the other hand, to provide a comprehensive review of their benefits and limitations when applied in the wild to either permissioned or permissionless blockchain frameworks.

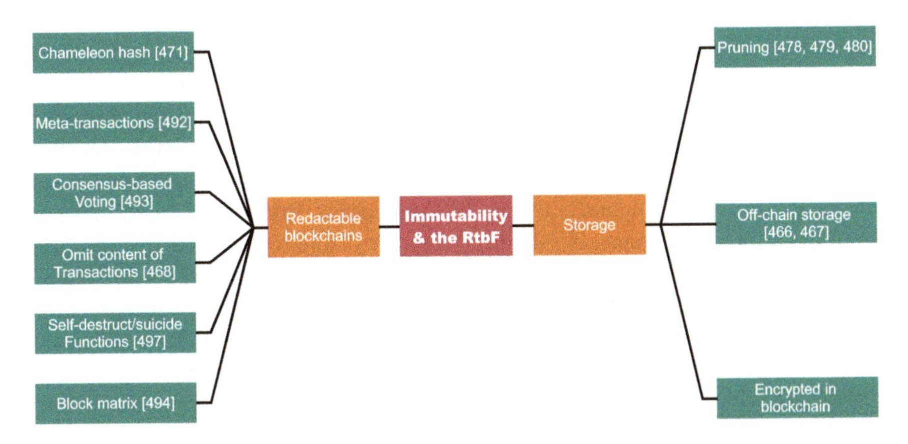

Fig. 7.1 An overview of solutions for balancing immutability and the right to be forgotten

7.4.1 Bypassing Blockchain's Immutability

A common workaround suggested throughout the literature for aligning blockchains with the GDPR privacy requirements is the use of blockchains only for storing a timestamp and a hash that points to the actual information held off-chain [33, 34]. Therefore, when information needs to be amended or deleted only the fact that the specific content version existed at a given point in time will remain in the blockchain. Bearing in mind that by using the stored hash alone the original content cannot be reconstructed, this workaround seems to resolve the blockchain's immutability collision with the RtbF rather elegantly. Indeed, off-chaining techniques so far are considered to be key tools in the engineering of blockchain-based application as they present significant benefits, such as reduced blockchain data storage requirements, and hence fewer scalability issues, as well as GDPR compliance [17, 33]. As a matter of fact, a recently conducted study [19] among experts concluded that blockchains could be indeed compliant with the privacy by design principles of the GDPR, and consequently with the RtbF, by employing these kinds of off-chaining techniques. On the downside, however, these techniques move the responsibility of robust, distributed data storage to other protocols like the IPFS [35], while they introduce complexity and additional delays. Furthermore, they have been criticized for decreasing blockchains' security by introducing more attack vectors [36, 37]. But most importantly, these solutions do not avoid the burden of having to remove the hash pointers from the blockchain since hashed data are pseudonymous, not anonymous, and therefore need to be protected as well [24]. For instance, hashed data may reveal sensitive personal information either when combined with other available information or when they are subject to dictionary attacks.

Another alternative solution for complying blockchains with the RtbF is to have the data stored in the blockchain in an encrypted form, and when the user asks to delete personal information, forgetting or deleting the encryption key will make the data inaccessible, i.e. no retrievable. Although some experts argue that, in the case of the blockchain, inaccessibility equals deletion, this is not the opinion of the data protection authorities such as the French CNIL which explained that, strictly speaking, this approach is not an actual erasure [26]. Another limitation of this solution stems from the difficulty in managing the decryption keys among many parties that need access to the data. Furthermore, there is always the case that personal data to become unreadable or available to everyone when the key is either lost or becomes accidentally known [23, 38, 39]. Taken into account that data shall remain encrypted across their life cycle, a further limitation derives from the rapid advancements in quantum computing which, according to experts, is going to break most encryption schemes used nowadays [40–43]. To avoid information be susceptible to decryption once quantum computers become available, sensitive data need to be protected in the long term by using symmetric algorithms with long key lengths. However, such a choice would have a severe impact on the storage requirements of the designed blockchain systems. Unless fully homomorphic encryption or some form of malleable encryption schemes is used, the processing of these data will also be impossible. But even then, the extra burden of processing and querying encrypted data would have a severe impact on the performance of the blockchain system [44].

Blockchain pruning is also proposed as a way to remove data from blockchains. While pruning has been originally discussed by Nakamoto in the bitcoin paper back in 2009 [45], it was not until 2015 that this feature was implemented in bitcoin [46]. In blockchain pruning, old transactions and blocks are deleted after a predefined amount of time, whereas old block headers containing the hashed version of the removed block data are maintained to ensure the integrity and security of the blockchain. While originally pruning aimed at compressing the blockchain size on the assumption that historical data are not required, it is argued that it can also offer an increased level of user privacy since old transactions might not be locatable. Accordingly, it can serve regulatory requirements allowing the old transactions to be forgotten from the network [17, 47, 48]. In this respect, a cryptocurrency scheme called the "mini-blockchain" has been proposed as a pruning alternative to current blockchain implementations [49]. The proposed scheme eliminates the need for a full blockchain by unlinking transactions, and therefore it allows all transactions to be discarded after a safe amount of time has elapsed. Obviously, when nodes discard the old blocks, they do not discard the block headers which are stored in a separate "proof-chain" to maintain the long term blockchain history. Although blockchain pruning meets scalability and privacy requirements, it has been argued that it does so at the expense of the security since, even when old block headers are maintained, truncating blockchain's history yields to a decreased security [17]. In addition, frequent pruning adds an expensive overhead that may result in further inconsistencies and scalability issues when blockchain's state is verified. Pruning has also been criticized for its weak enforceability as there is not any guarantee that all nodes will choose not to store the full chain. Nonetheless, it has been foreseen that pruning may be an appropriate solution for permissioned blockchain frameworks where the operating environment is more easily controlled and adjusted [50]. Yet, the idea of pruning in public blockchains remains controversial, and it is nowadays an active field of research [51].

7.4.2 Removing Blockchain's Immutability

Much has been written on the advantages and disadvantages of having a mutable blockchain, i.e. a blockchain whose content can be edited or deleted. While for business technocrats, the idea seems rather reasonable as it may adapt blockchains to enterprises' requirements and constraints, for crypto proponents the idea seems repulsive as it eradicates the blockchain's append-only and censorship-resistance nature. Indeed, the very first purpose of introducing the blockchain, which was to provide decentralized trust on committed transactions, is being cancelled out once the ledger becomes editable. Furthermore, an editable blockchain brings the question by whom should be edited and under which circumstances. Despite the arguments on both sides of the debate, the technical implementation for introducing mutability to blockchains is not an easy task. Technologically speaking, the research on removing blockchain's immutability while preserving security is still in infant stages. Nonetheless, some interesting cryptographic and innovative proposals towards this end are discussed below.

Reversing transactions in fraudulent or exceptional cases was discussed among bitcoin developers and blockchain thinkers even from the early days of cryptocurrency boom [52]. However, since bitcoin was built by design as being immutable for security purposes, crypto supporters were never in favour of such an option. In fact, editing the bitcoin blockchain has been met with strong criticism since bitcoin's immutability is regarded as a valuable feature. Reversecoin however, was the first altcoin that attempted to reverse transactions within a timeout period [53]. Its idea was to enable users to seamlessly transact with their online wallets and fall back to an offline wallet if their online account gets hacked [54]. Reversecoin worked by setting two different kinds of accounts: Standard Accounts, which are like bitcoin accounts; and Vault Accounts, which are like bank savings accounts. Each vault account has a configurable timeout and is backed by two key pairs, one online and one offline. Only the online key pair is needed to transfer coins from a vault, and the resulting transactions are confirmed after they live in blockchain for the timeout period. During this period, one can reverse those transactions by using the offline key pair and restore the coins in case the transaction originated by a malicious user. Additionally, all reverted transactions remained untouched in the blockchain history so can be publicly viewed. Unfortunately, although reversecoin's original idea was rather appealing, the project did not enjoy widespread acceptance.

The first technical proposal that actually challenged blockchain's immutability is the one published by Ateniese et al. [38]. The authors proposed the replacement of the hash function that connects each block to the previous one with an evolution of the standard chameleon hash. A chameleon hash is a cryptographic hash function that contains a trapdoor, and the knowledge of this trapdoor allows collisions to be generated efficiently [55]. While in a standard chameleon hash collisions must be kept private since the trapdoor can be extracted from a single collision, in the proposed improved design it is safe to reveal any number of collisions. With the knowledge of the trapdoor key, it is possible to efficiently find collisions and thus replace the content of the blocks. Thereby, knowing the key, any redaction of the blockchain is possible, including deletion, modification, and insertion of any number of blocks. The proposed system also leaves an immutable "scar" to indicate when any blocks have been altered, maintaining thus auditability and transparency. Researchers' main idea was to have the trapdoor key be secretly shared among some fixed set of users that are in charge of redacting the blockchain content in specific and exceptional circumstances. For example, the key could be in the hands of miners, a centralized auditor, or shares of the key could be distributed among several authorities, so that unanimous agreement must be reached to make any changes. However, the question of how these keys should be appropriately protected and managed remains open.

Unavoidably, the announcement of the first redactable blockchain was met with widespread derision while provoked a lot of agitation and skepticism among blockchain believers and cryptocurrency advocates who even argued that an editable blockchain is actually similar to a database [56–59]. They were further claiming that having to trust a set of specific participating authorities, such as banks, to edit the blockchain contents invalidates the decentralized nature of blockchains and defeats the very benefit of this technology [17, 60]. In addition, they argued that a redactable blockchain opens up the financial systems to possible fraudulent activities because the disclosure of the trapdoor key makes the blockchain vulnerable to malicious attacks and decreases its security [61]. Despite the criticism, the authors teamed up with Accenture, a big consulting firm, to develop a prototype adapted and refined for permissioned environments based on Hyperledger. Notwithstanding the author's argument that the solution is compatible with current blockchain frameworks, both permissionless and permissioned, sharing the key needed to edit a blockchain to a finite number of trusted nodes renders the solution suitable only for permissioned settings. However, as stated in [60], in permissioned blockchains mutations can be performed much more easily based on a voting process, albeit less optimized in terms of performance. Nevertheless, the proposed solution can be used temporarily as a way to fix any irregularities in a permissioned blockchain. In addition, the technique can be exploited to address the scalability problem of blockchains because chameleon hashes can be used to compress the ledger and to consolidate the transactions.

Another technical solution for forgetting data stored in blockchains is proposed in [62] where a mutable blockchain that enables the deletion and modification of blockchain content is described. The proposed design leverages the consensus mechanisms of traditional blockchains to vote on alternate versions of blockchain history. It does so through the introduction of mutable transactions which represent transactions sets that contain various possible versions of transactions. In a transaction set, only one of the transactions is specified as active, while all the others are inactive alternatives. All modifications are performed using transactions of a special type, meta-transactions, which are issued by users or smart contracts and are verified by validators. Mutations are also subject to access control policies specified by the transaction senders. These policies define who, and under which circumstances, is allowed to trigger mutations or to add additional versions of data records, and validators verify their conditions. To hide alternative history versions, the blockchain relies on encryption: all possible transaction versions are encrypted using transaction-specific keys whereas only the decryption keys for the active records are made available. To adapt the setting to the constraints of permissionless blockchains, the authors use a secret sharing scheme to split the transaction-specific keys into shares and distribute those shares among the validators, which can only reconstruct the entire key if a sufficient number of shares are collected. However, as the authors state, this scheme adds significant performance overhead and limits the verification enforcement of some transaction properties. Additionally, while the proposed blockchain offers solutions to the patching of vulnerable smart contracts and the elimination of abusive content from blockchains, it also presents some limitations that hinder its wide acceptance as a forgetting mechanism in permissionless settings. For instance, once an active transaction becomes inactive due to mutation, and therefore its decryption key is not served anymore by validators, local copies of keys may remain stored locally by clients. As a result, the reconstruction of an inactive, i.e. "forgotten", record is still possible.

Criticizing the above proposal for allowing a malicious user in a public blockchain to simply not include a mutation for his transaction, or even to set a policy where only he himself can mutate the transaction, the authors of [63] present a redactable blockchain that does not rely on heavy cryptographic tools and is suitable for permissionless settings. Its protocol uses a consensus-based voting based on a PoW and is parameterized by a policy that dictates the requirements and constraints for the redactions. Any user can propose the edit operations but they are only performed if approved by the blockchain policy (e.g., voted by the majority). Moreover, the protocol offers accountability for edit operations as any edit in the chain can be publicly verified. Nonetheless, although the proof-of-concept implementation of the proposed scheme presents only a tiny overhead in the chain validation when compared to an immutable one, the proposed permissionless blockchain operates on the assumption that the majority of the miners in the network are honest, and they behave rationally when they vote to either accept or reject the edit requests.

In [36] a memory flexible blockchain framework tailored towards IoT networks is presented. The framework allows users to modify, compress, or completely remove their transactions from blockchains while it preserves transactions' consistency. This is achieved by computing the hash of the block over the hashes of its constituted transactions and not of their contents, thereby allowing a transaction to be removed from a block without impacting the hash consistency checks. In particular, for each transaction stored in the blockchain, a specific value is calculated as the signed hash of a secret only the entity generating the transaction knows. To remove a stored transaction, the user has to prove that it has previously generated that transaction by including in the remove transaction the hashes used to generate the secret of the transaction to be removed and the encrypted form of the hashed secret using her public key. When a transaction is removed, while its content is removed from the blockchain, the hash of its content and the hash of its preceding transaction remain stored in the blockchain to ensure blockchain consistency and auditability. To facilitate the removal process, multiple agents are introduced to reduce the packet and processing overhead associated with multiple memory optimization methods used. Each agent is identified by a unique public key which is certified by a Certificate Authority (CA) to verify its identity. Moreover, for maintaining consistency among transactions and for auditing purposes, a shared read-only central database known as a blackboard and managed centrally by a Blackboard Manager Agent (BMA) is employed. Multiple replications of the blackboard exist to reduce the risk of single point of failure and to ensure scalability. Overall, the proposed framework provides a solid technical framework suitable for compressing, modifying and removing transaction data from blockchains in IoT environments. Yet, since it relies heavily on centralized entities (CA and BMA) for the management of its key functionalities (agents and blackboard), it significantly deviates from a fully decentralized solution.

In another research, the problem of preserving hash-based integrity when deleting transactions from blockchains is tackled [64]. The author describes a data structure, a block matrix, and an algorithm that allow the safe deletion of arbitrary records while preserving hash-based integrity assurance that other blocks remain unchanged. However, the solution has been thus far focused only to permissioned blockchains to ensure their transaction integrity and their compliance with the erasing requirements of the RtbF [65]. Nevertheless, the idea appears rather appealing as it delves into a core blockchain element, its data structure.

Similarly to blockchain transaction data and contrary to traditional distributed applications that can be patched when bugs are detected, smart contracts living on the blockchain are also irreversible and immutable [66]. In other words, once smart contracts' code is migrated to the blockchain network there is no way to patch bugs or alter their functionalities. Smart contracts are not removed from the blockchain when their use has come to an end. Instead, they are part of the history of the blockchain and probably retained by most nodes. Even when developers think in advance a way to disable them manually, by inserting ad-hoc code in the contracts, or automatically, by calling self-destruct or suicide functions, the smart contracts are still present but unresponsive [67, 68]. Yet, smart contracts' immutability refers only to their actual code and not to their state which is mostly set from the state of their variables and functions. In fact, in ethereum network, while variables' state can change freely, the history of storage variables in contracts is permanently stored. Furthermore, the functions in the contracts' code are immutable once they are deployed to the blockchain. This immutability is exploited by decentralized applications (DApps) to store some data persistently, and in some cases to certify data ownership and provenance, e.g. to write the hash of a document on the blockchain so that they can prove document existence and integrity [68]. However, due to their immutable nature of smart contracts, their correctness has been identified as a critical factor for their proper and safe behaviour [66, 69]. Furthermore, acknowledging that, in contrast to their analogue counterparts, smart contracts' immutability does not allow traditional tools of contract law for termination, rescission, modification and reformation, to be applied successfully to smart contracts, researchers are arguing for a new set of standards to alter and undo smart contracts in order to ensure that the traditional tools achieve their original (contract law) goals when applied to the blockchain technology [70].

7.5 The Controversy

Over the last years, considerable prominence has been given on the controversy over the blockchain's immutability, mainly due to the adoption of the GDPR and, hence, of the RtbF which foresees the retroactive erasure of personal data upon request and from all available places to which they have been disseminated. According to blockchain immutability, on the other hand, which as described in 5.3.1.3 is comprised by the tamper-evident and tamper-resistant properties of blockchain, tampering with data stored in blockchains is nearly impossible. Consequently, immutability facilitates the single, globally accepted view of events among non-trusted participants. In other words, immutability supports the possibility of decentralized trust in inherently trust-less interactions. Therefore, for cryptocurrency activists and blockchain proponents even simply questioning the immutable nature of blockchain is tantamount to heresy

[57]. They argue that an editable blockchain is just a database since the basic argument for a blockchain is that it is a tamper-proof immutable ledger that provides an uncensored truth. In that respect, the RtbF is regarded as an obstacle to the widespread adoption of blockchain technology. On the opposite side, privacy advocates look upon blockchains' immutability as a risk to people's data protection and privacy rights. For enterprise technocrats, however, incorporating limited mutability within permissioned blockchain systems, subject to certain conditions, can strike the right balance between preserving blockchain's key features and adapting it for real-world requirements [71].

In view of the above, the recent advancements on introducing mutability, based on strict, pre-approved rules, appeals both to regulators and to enterprises [21]. As a matter of fact, the number of public authorities that are already exploring the use of blockchain for their administration and services is rising [72–74]. In 2017, DG TAXUD, the EU General Directorate responsible for EU policies on taxation and customs, started exploring blockchain technology within the customs domain [75], while in 2018 21 EU Member States plus Norway agreed to sign a declaration creating the European Blockchain Partnership (EBP) and to cooperate in the establishment of a European Blockchain Services Infrastructure (EBSI) that will support the delivery of cross-border digital public services [76]. At about the same time, the European Commission with the support of the European Parliament launched the EU Blockchain Observatory and Forum with the purpose to encourage governments, industry and citizens to benefit from blockchain opportunities [77]. Similarly, the OECD has begun investigating the benefits and risks of blockchain for economies and societies [78, 79], while the UN is gradually embracing blockchain technology [80]. In the banking sector, the use of digital currencies based on blockchain technology is progressing rapidly as many major banks have already announced blockchain projects to build new digital currencies [81–83]. Enterprises also consistently engage and invest in the blockchain technology [84–87].

Notwithstanding these global initiatives towards a blockchain-enabled era, the blockchain's mass-market adoption is not expected any time soon [88]. In particular, experts believe that blockchain is now where the web was in 1994 [89]. According to Gartner, while blockchain is one of the emerged trends in 2018, it is expected to reach a healthy and stable plateau at least in five to ten years [90]. Despite blockchain's slow integration into real-life applications, the extent to which blockchain's incompatibility with data protection and privacy rights occupies the interest of scientists and businesses is remarkable. In that respect, we believe that discussing, analysing and resolving disputed areas of blockchain technology, such as immutability, will be proved valuable both to industry and to academia.

References

1. European Union, Regulation (EU) 2016/679 of the European Parliament and of the Council of 27 April 2016 on the protection of natural persons with regard to the processing of personal data and on the free movement of such data, and repealing Directive 95/46/EC (General Data Protection Regulation), Official Journal of the European Union, L 119 (4 May 2016) (2016), pp. 1–88
2. E. Politou, E. Alepis, C. Patsakis, Forgetting personal data and revoking consent under the GDPR: Challenges and proposed solutions. J. Cybersecurity **4**(1):tyy001 (2018)
3. K. Christidis, M. Devetsikiotis, Blockchains and smart contracts for the internet of things. IEEE Access **4**, 2292–2303 (2016)
4. A. Biryukov, D. Khovratovich, I. Pustogarov, Deanonymisation of clients in bitcoin p2p network, in *Proceedings of the 2014 ACM SIGSAC Conference on Computer and Communications Security* (ACM, 2014), pp. 15–29
5. S. Goldfeder, H. Kalodner, D. Reisman, A. Narayanan, When the cookie meets the blockchain: privacy risks of web payments via cryptocurrencies. Proc. Priv. Enhancing Technol. **4**, 179–199 (2018)
6. S. Meiklejohn, Top ten obstacles along distributed ledgers path to adoption. IEEE Secur. Priv. **16**(4), 13–19 (2018)
7. S. Meiklejohn, M. Pomarole, G. Jordan, K. Levchenko, D. McCoy, G.M. Voelker, S. Savage, A fistful of bitcoins: characterizing payments among men with no names, in *Proceedings of the 2013 Conference on Internet Measurement Conference* (ACM, 2013), pp. 127–140
8. D.F. Primavera, The interplay between decentralization and privacy: the case of blockchain technologies. J. Peer Prod. **9** (2016)
9. E. Androulaki, G.O. Karame, M. Roeschlin, T. Scherer, S. Capkun, Evaluating user privacy in bitcoin, in *International Conference on Financial Cryptography and Data Security* Springer, Berlin (2017), pp. 34–51
10. M. Fleder, M.S. Kester, S. Pillai, Bitcoin transaction graph analysis (2015). arXiv preprint arXiv:150201657
11. F. Tschorsch, B. Scheuermann, Bitcoin and beyond: a technical survey on decentralized digital currencies. IEEE Commun. Surv. Tutorials **18**(3), 2084–2123 (2016)
12. D. Gerard, Blockchain identity: Cambridge Analytica, but on the blockchain (2018). https://davidgerard.co.uk/blockchain/2018/03/22/blockchain-identity-cambridge-analytica-but-on-the-blockchain/
13. V. Buterin, Privacy on the blockchain (2016). https://blog.ethereum.org/2016/01/15/privacy-on-the-blockchain/
14. D. Hopwood, S. Bowe, T. Hornby, N. Wilcox, *Zcash Protocol Specification* (Tech. Rep. 2016–110 Zerocoin Electric Coin Company, Tech Rep, 2016)
15. G. Zyskind, O. Nathan, A. Pentland, Enigma: decentralized computation platform with guaranteed privacy (2015). arXiv preprint arXiv:150603471
16. S. Meiklejohn, R. Mercer, Möbius: Trustless tumbling for transaction privacy. Proc. Priv. Enhancing Technol. **2**, 105–121 (2018)
17. M. Finck, Blockchains and data protection in the European Union. Eur Data Prot. Law Rev. **4**, 17 (2018)
18. R. Thurimella, Y. Aahlad, The hitchhiker's guide to blockchains: a trust based taxonomy (2018). https://wandisco.com/assets/whitepapers/the-hitchhikers-guide-to-blockchains.pdf
19. S. Schwerin, Blockchain and privacy protection in the case of the European General Data Protection Regulation (GDPR): a delphi study. The JBBA **1**(1), 3554 (2018)
20. L.D. Ibáñez, K. O'Hara, E. Simperl, On blockchains and the General Data Protection Regulation (2018). https://eprints.soton.ac.uk/id/eprint/422879
21. S. Sater, Blockchain and the European Union's General Data Protection Regulation: A Chance to Harmonize International Data Flows. Available at SSRN 3080987 (2017)
22. D.A. Zetzsche, R.P. Buckley, D.W. Arner, The distributed liability of distributed ledgers: Legal risks of blockchain. Univ. Ill. Law Rev. 1361 (2018)

23. E. Politou, A. Michota, E. Alepis, M. Pocs, C. Patsakis, Backups and the right to be forgotten in the GDPR: an uneasy relationship. Comput. Law Secur. Rev. **34**(6), 1247–1257 (2018)
24. Article 29 Data Protection Working Party (2011) “Opinion 15/2011 on the definition of consent” WP 187. https://ec.europa.eu/justice/article-29/documentation/opinion-recommendation/files/2011/wp187_en.pdf
25. Commission Nationale de l’Informatique et des Libertes̃ (CNIL) (2018) Blockchain, solutions for a responsible use of the blockchain in the context of personal data. https://www.cnil.fr/en/blockchain-and-gdpr-solutions-responsible-use-blockchain-context-personal-data
26. F. Martin-Bariteau, Blockchain and the European Union General Data Protection Regulation: The CNIL’s Perspective. Blckchn ca Working Paper Series 1 (2018)
27. European Data Protection Supervisor (EDPS), Opinion 5/2018, Preliminary Opinion on privacy by design (2018). https://edps.europa.eu/sites/edp/files/publication/18-05-31_preliminary_opinion_on_privacy_by_design_en_0.pdf
28. European Parliament, Resolution of 3 October 2018 on distributed ledger technologies and blockchains: building trust with disintermediation (2017/2772(RSP)) (2018). http://www.europarl.europa.eu/sides/getDoc.do?type=TA&reference=P8-TA-2018-0373&language=EN
29. A. Toth, Will GDPR block Blockchain? (2018) https://www.weforum.org/agenda/2018/05/will-gdpr-block-blockchain/
30. C. Wirth, M. Kolain, Privacy by blockchain design: a blockchain-enabled GDPR-compliant approach for handling personal data, in *Proceedings of 1st ERCIM Blockchain Workshop 2018, European Society for Socially Embedded Technologies (EUSSET)* (2018)
31. T. Swanson, Consensus-as-a-service: a brief report on the emergence of permissioned, distributed ledger systems, 2015 (2015)
32. D. Treat, Accenture: absolute immutability will slow blockchain progress (2016). https://www.coindesk.com/absolute-immutability-will-slow-permissioned-blockchain-progress
33. J. Eberhardt, S. Tai, On or off the blockchain? insights on off-chaining computation and data, in *European Conference on Service-Oriented and Cloud Computing* (Springer, Berlin, 2017), pp. 3–15
34. E. García-Barriocanal, S. Sánchez-Alonso, M.A. Sicilia, Deploying metadata on blockchain technologies, in *Research Conference on Metadata and Semantics Research* (Springer, Berlin, 2017), pp. 38–49
35. (2021) IPFS, InterPlanetary File System. https://ipfs.io/
36. A. Dorri, S.S. Kanhere, R. Jurdak, MOF-BC: a memory optimized and flexible blockchain for large scale networks. Future Genera. Comput. Syst. **92**, 357–373 (2019)
37. A. Van Humbeeck, The Blockchain-GDPR Paradox (2017). https://medium.com/wearetheledger/the-blockchain-gdpr-paradox-fc51e663d047
38. G. Ateniese, B. Magri, D. Venturi, E. Andrade, Redactable blockchain–or–rewriting history in bitcoin and friends, in *2017 IEEE European Symposium on Security and Privacy (EuroS&P)* (IEEE, 2017), pp. 111–126
39. Open Data Institute, Applying blockchain technology in global data infrastructure (2016). https://theodi.org/article/applying-blockchain-technology-in-global-data-infrastructure
40. (2018) Quantum computers will break the encryption that protects the internet. https://www.economist.com/science-and-technology/2018/10/20/quantum-computers-will-break-the-encryption-that-protects-the-internet
41. D. Aggarwal, G.K. Brennen, T. Lee, M. Santha, M. Tomamichel, Quantum attacks on bitcoin, and how to protect against them (2017). arXiv preprint arXiv:171010377
42. M. Dunjic, Blockchain immutability ... blessing or curse? (2018) https://www.finextra.com/blogposting/15419/blockchain-immutability--blessing-or-curse
43. F. Lardinois, IBM unveils its first commercial quantum computer (2019). https://techcrunch.com/2019/01/08/ibm-unveils-its-first-commercial-quantum-computer/?guccounter=1
44. H.T. Vo, A. Kundu, M.K. Mohania, Research directions in blockchain data management and analytics, in EDBT (2018), pp. 445–448
45. S. Nakamoto et al., Bitcoin: A peer-to-peer electronic cash system (2008)

46. (2015) Bitcoin Core version 0.11.0 Release Notes: Block file pruning. https://github.com/bitcoin/bitcoin/blob/v0.11.0/doc/release-notes.md/#block-file-pruning
47. S. Farshid, A. Reitz, P. Roßbach, Design of a forgetting blockchain: a possible way to accomplish GDPR compatibility, in *Proceedings of the 52nd Hawaii International Conference on System Sciences* (2016)
48. R. Géraud, D. Naccache, R. Roşie, Twisting lattice and graph techniques to compress transactional ledgers, in *International Conference on Security and Privacy in Communication Systems* (Springer, Berlin, 2017), pp. 108–127
49. J. Bruce, The mini-blockchain scheme (2014). http://cryptonite.info
50. E. Palm, Implications and impact of blockchain transaction pruning (2017)
51. (2018) Ethereum chain pruning for long term 1.0 scalability and viability. https://ethereum-magicians.org/t/ethereum-chain-pruning-for-long-term-1-0-scalability-and-viability/2074/3
52. (2011) Can a bitcoin transaction be reversed? https://bitcoin.stackexchange.com/questions/197/can-a-bitcoin-transaction-be-reversed
53. (2021) Reversecoin. http://www.reversecoin.org/
54. (2014) Reversecoin—world's first cryptocurrency with reversible transactions. https://bitcoinist.com/reversecoin-worlds-first-cryptocurrency-reversible-transactions/
55. H. Krawczyk, R. Tabin, Chameleon signatures, in *Proceedings of the Network and Distributed System Security Symposium, NDSS 2000, San Diego, California, USA, The Internet Society, 2000*. http://www.isoc.org/isoc/conferences/ndss/2000/proceedings/042.pdf
56. J. Kelly, Accenture breaks blockchain taboo with editing system (2016). https://www.reuters.com/article/us-tech-blockchain-accenture-idUSKCN11Q1S2
57. R. Lumb, Downside of bitcoin: a ledger that can't be corrected (2016). https://www.nytimes.com/2016/09/10/business/dealbook/downside-of-virtual-currencies-a-ledger-that-cant-be-corrected.html
58. J.J. Roberts, Why accenture's plan to 'edit' the blockchain is a big deal (2016). http://fortune.com/2016/09/20/accenture-blockchain/
59. D. Roua, A patent for "an editable blockchain"? well, if it's editable, it's not a blockchain, dude, it's a database (2018). https://steemit.com/blockchain/@dragosroua/a-patent-for-an-editable-blockchain-well-if-it-s-editable-it-s-not-a-blockchain-dude-it-s-a-database
60. G. Greenspan, The blockchain immutability myth (2017). https://www.multichain.com/blog/2017/05/blockchain-immutability-myth/
61. J. Althauser, Accenture secures patent for its "editable blockchain" technology (2017). https://cointelegraph.com/news/accenture-secures-patent-for-its-editable-blockchain-technology
62. I. Puddu, A. Dmitrienko, S. Capkun, μchain: how to forget without hard forks. IACR Cryptology ePrint Archive **2017**, 106 (2017)
63. D. Deuber, B. Magri, S. Thyagarajan, Redactable blockchain in the permissionless setting, in *IEEE Symposium on Security and Privacy (SP)* (IEEE Computer Society, Los Alamitos, CA, USA, 2019)
64. R. Kuhn, A data structure for integrity protection with erasure capability (draft) (2018). https://csrc.nist.gov/publications/detail/white-paper/2018/05/31/data-structure-for-integrity-protection-with-erasure-capability/draft
65. R. Kuhn, Supporting GDPR Requirements and Integrity in Distributed Ledger Systems (2018). https://blockchain.ieee.org/images/files/pdf/20180918-supporting-gdpr-requirements-and-integrity-in-distributed-ledger-systems_-_r-kuhn.pdf
66. L. Luu, D.H. Chu, H. Olickel, P. Saxena, A. Hobor, Making smart contracts smarter, in *Proceedings of the 2016 ACM SIGSAC Conference on Computer and Communications Security* (ACM, 2016), pp. 254–269
67. (2021) Solidity v0.5.6, Introduction to Smart Contracts. https://solidity.readthedocs.io/en/v0.5.6/introduction-to-smart-contracts.html
68. M. Bartoletti, L. Pompianu, An empirical analysis of smart contracts: platforms, applications, and design patterns, in *International Conference on Financial Cryptography and Data Security* (Springer, Berlin, 2017), pp. 494–509

69. K. Bhargavan, A. Delignat-Lavaud, C. Fournet, A. Gollamudi, G. Gonthier, N. Kobeissi, N. Kulatova, A. Rastogi, T. Sibut-Pinote, N. Swamy et al., Formal verification of smart contracts: Short paper, in *Proceedings of the 2016 ACM Workshop on Programming Languages and Analysis for Security* (ACM, 2016), pp. 91–96
70. B. Marino, A. Juels, Setting standards for altering and undoing smart contracts, in International Symposium on Rules and Rule Markup Languages for the Semantic Web (Springer, Berlin, 2016), pp. 151–166
71. S. Das, Accenture to unveil 'editable' blockchain prototype (2016). https://www.ccn.com/accenture-unveil-editable-blockchain-prototype
72. B. Allison,Guardtime secures over a million estonian healthcare records on the blockchain (2016). http://www.openhealthnews.com/news-clipping/2016-03-03/guardtime-secures-over-million-estonian-healthcare-records-blockchain
73. R. Kalvapalle, Canada trialing use of ethereum blockchain to enhance transparency in govt funding (2018). https://globalnews.ca/news/3977745/ethereum-blockchain-canada-nrc/
74. S. Ølnes, A. Jansen, Blockchain technology as infrastructure in public sector: An analytical framework, in *Proceedings of the 19th Annual International Conference on Digital Government Research: Governance in the Data Age* (ACM, 2018), p. 77
75. Z. Saadaoui, Blockchain@TAXUD experience (2018). https://www.wto.org/english/res_e/reser_e/session_2c_4_zahouani_saadaoui_dg_taxud_blockchain_v1.0.pdf
76. European Commission, European countries join Blockchain Partnership (2018b). https://ec.europa.eu/digital-single-market/en/news/european-countries-join-blockchain-partnership
77. European Commission, European Commission launches the EU Blockchain Observatory and Forum (2018a). http://europa.eu/rapid/press-release_IP-18-521_en.htm
78. (2018) Blockchain and distributed ledger technology. http://www.oecd.org/finance/blockchain/
79. G. Medcraft, The OECD and the Blockchain Revolution (2018). https://www.oecd.org/parliamentarians/meetings/meeting-on-the-road-london-april-2018/The-OECD-and-the-Blockchain-Revolution-Presentation-by-Greg-Medcraft-delivered-on-29-March-2018.pdf
80. United Nations, Usage of Blockchain in the UN System (2017). https://unite.un.org/sites/unite.un.org/files/session_3_b_blockchain_un_initiatives_final.pdf
81. S. Das, China's Central Bank Completes Digital Currency Trial on a Blockchain (2017). https://www.ccn.com/chinas-central-bank-completes-digital-currency-trial-blockchain
82. S. Higgins, Spanish Banks Form New Blockchain Consortium (2017). https://www.coindesk.com/spanish-banks-form-new-blockchain-consortium
83. S. O'Neal, Central Bank-Issued Digital Currencies: Why Governments May (or May Not) Need Them (2018). https://cointelegraph.com/news/central-bank-issued-digital-currencies-why-governments-may-or-may-not-need-them
84. (2018) Air France-KLM partners with Winding Tree to strengthen innovation in the travel industry using Blockchain technology. https://www.airfranceklm.com/en/news/air-france-klm-partners-winding-tree-strengthen-innovation-travel-industry-using-blockchain
85. ConsenSys, Deloitte: 95% of Companies Surveyed Are Investing in Blockchain Tech (2019). https://media.consensys.net/deloitte-95-of-companies-surveyed-are-investing-in-blockchain-tech-5566f4942b5d
86. T. Donovan Barnett, These 20 Companies are Placing Big Bets on Blockchain Technology (2018). https://interestingengineering.com/these-20-companies-are-placing-big-bets-on-blockchain-technology
87. M. Marquit, More Mainstream Companies Invest in Blockchain (2017). https://due.com/blog/mainstream-companies-blockchain/
88. Y. Vilner, No More Hype: Time To Separate Crypto From Blockchain Technology (2018). https://www.forbes.com/sites/yoavvilner/2018/11/14/no-more-hype-time-to-separate-crypto-from-blockchain-technology/#1d5ea73b171c

89. D. Bradbury, Hyperledger 3 years later: That's the sound of the devs... working on the chain ga-a-ang. But is anyone actually using it? https://www.theregister.co.uk/2018/01/02/hyperledger_at_three/
90. K. Panetta, 5 Trends Emerge in the Gartner Hype Cycle for Emerging Technologies, 2018 (2018). https://www.gartner.com/smarterwithgartner/5-trends-emerge-in-gartner-hype-cycle-for-emerging-technologies-2018/

Chapter 8
Implementing Content Erasure in IPFS

Abstract The InterPlanetary File System (IPFS) is employed extensively nowadays by many blockchain projects to store personal data off-chain in order to comply with the Right to be Forgotten (RtbF) provision of the General Data Protection Regulation (GDPR). Nevertheless, upon an erasure request under the RtbF, the onus of removing the actual personal information moves to the IPFS protocol which—due to its decentralized nature—cannot guarantee data erasure across its entire network. Against this background, in this chapter we formalize a secure and anonymous protocol for delegated content erasure requests in the IPFS. The protocol can be easily integrated into the IPFS to disseminate an erasure request among all of its nodes and, eventually, to meet the RtbF requirements. Furthermore, the protocol complies with the primary principle of the IPFS to prevent censoring since the erasure is only allowed to the original content providers or their delegates. We demonstrate the efficacy of the proposed protocol by providing security proofs and experiments which confirm that the overhead introduced does not affect the overall performance of the IPFS.

8.1 Introduction

As the use of blockchains is generally prevented in applications that are processing personal information, decentralized p2p systems for file storage and sharing, and most particularly the IPFS, are employed extensively nowadays by many blockchain projects to store personal data off-chain to comply with the Right to be Forgotten (RtbF) of the General Data Protection Regulation (GDPR), the new regulatory regime for personal data protection in the EU. According to this practice, when a request for content erasure is to be carried out under the RtbF, the onus of removing the actual personal information moves to the IPFS protocol. Nevertheless, enforcing data erasure across the entire IPFS network is not actually possible, mainly due to its decentralized nature.

Notwithstanding the fact that the GDPR has not taken into account emerging decentralized technologies such as the blockchain, let alone the IPFS, when legislated the RtbF, its impact on such networks is rather substantial. Hence, the imple-

E. Politou et al., *Privacy and Data Protection Challenges in the Distributed Era*, Learning and Analytics in Intelligent Systems 26,
https://doi.org/10.1007/978-3-030-85443-0_8

mentation of a secure erasure mechanism based on delegates for handling content erasure requests within the IPFS would be the most constructive way towards aligning the IPFS with the GDPR. To that end, in this chapter we formally introduce an anonymous and secure protocol for content erasure requests in the IPFS. The proposed protocol can be smoothly integrated into the IPFS to distribute a request for erasure among all the IPFS nodes and, ultimately, to satisfy the erasure requirements foreseen in the RtbF. Furthermore, the protocol conforms to the primary principle of the IPFS to prevent censoring; therefore, erasure is only allowed to the original content providers or their delegates. We provide a formal definition and the security proofs, along with a set of experiments that prove the efficacy of the proposed protocol, and we demonstrate that the overhead introduced by our protocol does not affect the system's efficiency. Our experimental results exhibit a robust performance as the overall performance of the IPFS is not affected by the (average) times for generating the content-dependent keys and for disseminating the erasure requests.

In what follows, and before proceeding with the description of our protocol, we discuss the demand for off-chain solutions and we review the IPFS in terms of its privacy. Next, we describe how data erasure is currently managed by the IPFS and we discuss the need for its compliance with the RtbF. Last, we formally define our protocol for delegated content erasure in the IPFS and we provide the appropriate security and performance proofs.

8.2 Storing Off-Chain Personal Data in the IPFS

In accordance with our thorough analysis in Chap. 7, blockchains are considered to be in some respects incompatible with privacy and data protection obligations due to their transparent and immutable nature that requires data to be kept indefinitely and in plain sight. Consequently, personal information is generally avoided to be included in blockchain transactions [1]. As a matter of fact, uploading data to a public blockchain suffers both from scalability and privacy issues [1, 2]. In terms of scalability, putting large amounts of data in a blockchain transaction is not advisable due to the high cost involved as well as the latency problems introduced when full nodes need to download the entire ledger. In addition, the immutability of blockchains is found to be in conflict with the RtbF defined in the GDPR, according to which data controllers have to erase any personal information upon request and when certain conditions apply [1, 3, 4]. Although recent research efforts towards introducing restricted mutability in blockchains are indeed remarkable, editing or completely removing information stored in public blockchains is thus far practically impossible [1].

To overcome blockchain scalability and privacy issues while maintaining the benefits of decentralization, blockchains are commonly coupled with off-chain storage solutions. These involve storing the actual data outside the blockchain and keeping only the timestamps and pointers to these data in the blockchain. This is a favoured workaround when personal data are at stake. Yet, in order these workarounds to be considered as fully decentralized solutions they need to be based on decentralized p2p systems such as those described in Sect. 5.3.2. Among those systems, the most widely used nowadays is the IPFS which is used in combination with blockchains to store off-chain the actual files.

By storing the actual files containing personal information off-chain in the IPFS network, and maintaining in the blockchain only the hash pointers to those files, blockchain's compliance with the RtbF is considered satisfactory. Yet, the overall alignment with the GDPR is disputable since even the IPFS presents some limitations in terms of its privacy. For instance, it does not provide an object-level encryption yet. Nevertheless, while personal information can be always encrypted before added to the network, encryption alone could not be the base for maintaining the confidentiality of private information because encryption algorithms may become obsolete over time due to, e.g. the advances of quantum computing and computing power. Over and above that, it should be pointed out that the IPFS does not provide access control at the IPFS connection level to restrict untrusted peers from getting unauthorized data. Recognizing the need for secure sharing of private data, other open-source projects, like Peergos,[1], are building on top of the IPFS protocol to provide strong encryption and surveillance resistance to the IPFS network.

More importantly, however, storing the actual personal data in the IPFS network does not remove the burden of erasing them should the RtbF be raised. Instead, it is of paramount importance users to be able to control the dissemination of their own data within the IPFS network, especially considering the emergence of novel threats [5, 6]. However, the IPFS protocol layer has not anticipate for such a mechanism, mainly due to the impracticability of enforcing data erasure across all nodes in a decentralized public network. Nevertheless, given the adverse implications of non-complying with privacy and data protection rights, the alignment of the IPFS with the RtbF is considered highly critical.

8.3 Erasing Content in IPFS

While the IPFS is being widely advertised as the new "permanent web" where information remains always available regardless of single point of failure attacks or censorship take-downs, the term "permanent" should not be misinterpreted to be equivalent to the permanent storage and availability of the uploaded content. Instead, according to clarifications provided by the core development team, the term refers to the *permanent reference* of the content to which an IPFS link points.[2] As described in Sect. 5.3.2.1, this results from the content addressability property which ensures that all resources are uniquely and permanently addressed by their contents. On the other hand, the *permanent availability* of the resources in IPFS is cited as storage persistency and is managed by the functionality of pinning, which excludes an object and its children from being garbage collected within an IPFS node. The garbage

[1] https://github.com/Peergos/Peergos.

[2] https://discuss.ipfs.io/t/deleting-content/202.

collector runs frequently to delete any cached data that the IPFS node downloaded when accessing the network resources. Whether content added in the IPFS network is stored persistently or not depends exclusively on the users that choose to pin this content in order not to be garbage-collected after a given period of time. Otherwise, unpinned content is automatically garbage collected, provided that the garbage collector is enabled to run on a schedule or manually. Obviously, the more IPFS nodes are pinning a specific file, the easier and faster another node can get it.

However, when a file is garbage collected, i.e. deleted, from a node, this does not ensure that it has also been deleted from all the other nodes that had previously accessed and thereby cached that file. What's more, if a node has pinned it, the file remains permanently available in the IPFS network. Therefore, as long as someone is willing to continue spending energy, i.e. money, to maintain an object online, the corresponding content will be *permanently* available in IPFS. To promote data persistency in the IPFS, Filecoin—when implemented—will incentivize people to share their unused storage for pinning IPFS files. In a nutshell, a file is perpetually preserved in the IPFS network if there is at least one node that is actively sharing it (i.e. by pinning it), while it can only be completely removed from the network if its original provider as well as all the other hosts serving it delete it and its cached copies throughout the network expire.

While the IPFS provides the option of building private IPFS networks (using IPFS Clusters to coordinate connection among peers who share a secret key[3]) in which an entity controlling all peers can collective unpin and thereby delete a file from all the peers participating in the private network, completely removing content previously uploaded to the IPFS public network cannot be guaranteed.

8.4 The Requirement for Total Content Erasure

As already mentioned, IPFS is a decentralized public trustless network, since nodes do not have to trust each other, and with no single point of failure since there is not any entity in control of the flow (or the very existence) of the information within the network Nevertheless, the malevolent uses for promoting and disseminating illegal or copyright-protected content, or even cases of infringing on human rights such as the right to privacy, have not been considered by the original vision and design of the IPFS (and of the Internet in general). Consequently, there not any efficient procedures supported by the IPFS for totally erasing (and thereby stopping from being disseminated across its entire network) any illegal, personal or copyrighted content. Arguably, this deficiency might have adverse implications for the compliance of the IPFS with, at least, the data protection and privacy laws.

Acknowledging this lack, IPFS intents to support blocklists, i.e. lists of illegal content that needs to be blocked from the IPFS network. These blocklists are going to specify policies for content storage and distribution and therefore will allow sub-

[3] https://github.com/ipfs/go-ipfs/blob/master/docs/experimental-features.md#private-networks.

networks of peers to agree upon sets of content they would wish to censor. Still, blocklists, as they have been so far designed, present some limitations. First of all, they cannot be universally applied since what is illegal in one jurisdiction is not necessarily in others; e.g. consider political or even religious-related content. Moreover, updating, maintaining, and coordinating these lists by the IPFS gateways will be challenging given the high demand for censoring of links/content and the continuously increasing size of these lists. Besides, there is always the case of easily circumventing such blocklists since by changing only a bit of the unwanted file the corresponding hash changes while the actual information remains practically unaffected. On top of these, since the subscription to these blocklists will be most probably optional, nothing would prevent a node from not subscribing. Last but not least, there must be meticulous methods in place to diligently examine and add links/content to these blocklists in order not to violate any other legal rights, such as the freedom of speech. Most importantly, however, these blocklists fall short of meeting the data protection requirements anticipated by the GDPR, and in particular the RtbF which mandates the erasure of personal data published in the IPFS network under certain conditions.

Considering the above described inadequacy of the IPFS to comply with the RtbF erasing provisions, in the following sections we explore the RtbF in terms of its applicability on the IPFS protocol.

8.5 Towards Aligning IPFS with the RtbF

Harmonizing the IPFS with the RtbF is not an easy task. In fact, any efforts to impose an erasure request would most probably be pointless since IPFS is a trustless network, and as such its nodes can never be trusted to respect and implement a request for content erasure. A node can keep data for as long as it desires, and there is nothing to prevent this from happening as there is not any way to verify that data in the IPFS have been removed from the entire network. Above all, the IPFS—similarly to the HTTP—is actually a protocol, a base layer foundation which can be used by other systems to be built upon. As a result, enforcing data manipulation rules on it is realistically unfeasible since application-specific functionality is implemented in a higher architectural layer.

Despite this constraint, considering the extensive use of IPFS to store personal data and the strict data protection provisions of the GDPR it is highly recommended—if not urgently—the implementation of a type of erasure request in the IPFS protocol layer to disseminate the request across its entire network. According to Article 17(2) of the GDPR:

> the controller, taking account of available technology and the cost of implementation, shall take reasonable steps, including technical measures, to inform controllers which are processing the personal data that the data subject has requested the erasure by such controllers of any links to, or copy or replication of, those personal data.

Otherwise stated, regardless of the possibility of enforcing an erasure request, there should be at least a method to notify the request to all data controllers holding the personal data under consideration. Nevertheless, each IPFS node essentially acts as a data controller in GDPR terms since by definition each IPFS node fully controls its stored data, or in other words, an IPFS node determines on its own the means and the purposes for the processing of these data. Hence, according to the GDPR, each IPFS node should at least be able to "*take reasonable steps, including technical measures, to inform*" other IPFS nodes holding these data about the erasure request.

Based on the above analysis, it is evident that—even though it is not feasible to enforce data erasure across all the IPFS nodes—implementing a delegation mechanism for securely managing erasure requests in the protocol layer would be the most optimum and application-agnostic way towards complying the IPFS with the RtbF. Thereby, the onus of enforcing the actual erasure is moved to each individual IPFS node which is a data controller on its own. Beyond any doubt, integrating such kind of erasure mechanism based on delegates into the IPFS protocol it will endorse its GDPR compliance considerably, and consequently, it will add real value to its future adoption by applications processing personal data. In this regard, in the following section we present the security and technical details of implementing and integrating such a protocol into the IPFS, along with its formal definition and its proof-of-concept experimental results.

8.6 The Proposed Protocol

8.6.1 Assumptions and Desiderata

The proposed protocol aims at allowing a user to request the erasure of content that she has already shared and consequently it is found in other nodes of the IPFS network. For this purpose, we introduce a "*proof-of-ownership*" in the proposed protocol so that the content is associated anonymously with a secret that only a set of designated users have. Therefore, once someone submits an erasure request of the content with the corresponding secret, each node verifies that the secret matches the "proof-of-ownership" and propagates the request to the other nodes. The request may be repeated periodically to cater for nodes which may be offline at the time of the initial request.

In our protocol we consider that the content removal is performed on all nodes conforming to the protocol and storing the content along with the "proof-of-ownership". Our main motivation behind the introduction of "proof-of-ownership" is to maintain the censorship-resistant nature of the IPFS network. To this end, no one can arbitrarily request the erasure of any content , e.g. for content that was not submitted by herself. Therefore, while the current IPFS model supports sharing of content in the form of, e.g. a file c, we augment it, to support the sharing of tuples of the form (c, s) where s is an encrypted string which will be used afterwards to prove the ownership when an erasure request is made for a specific file c. Upon an erasure request, a content-dependent key k is sent to the network along with the hashed version of the file. The content-dependent key k derives from a master key that each user owns, preventing thus the hustle of having to remember different keys. Note that the proposed extension would allow users to continue sharing content without proofs of ownerships. The proofs are only appended if the user considers that she may once want to delete her shared files.

Apparently, extending the IPFS protocol to embrace the proof-of-ownership s and to accordingly support the manipulation of the corresponding content is a major divergence from its current design. Nevertheless, it is a feasible and practical alternative for allowing IPFS to manage erasure requests successfully. In what follows, we assume that we have honest nodes which comply with the protocol and they duly follow its rules.

8.6.2 Threat Model

Bearing in mind that the scope of this work is to allow the content erasure, the attacks that could be launched will target the erasure of a user's contents without her knowledge or consent. To achieve this, an adversary would attempt to extract the key, or forge the erasure request to achieve the removal of the content from all IPFS nodes. It is also assumed that attempts to bypass the protocol so that the content is not removed, e.g. not disseminating and not conforming to the erasure request, are beyond the scope of this work as they imply a node that does not comply with the protocol. Nodes of this kind are misbehaving and can be isolated from the rest of the network.

In our protocol, we assume probabilistic polynomial time (PPT) passive adversaries that are polynomially bounded and they do not have the ability to break the underlying cryptographic primitives used in the protocol, i.e. reverse hash functions or break any secure block cipher. We also assume that an adversary is able to monitor all the traffic exchanged within the protocol execution. Furthermore, active attacks are not taken into account since we assume that the messages exchanged while the protocol is executed are authenticated, and their integrity is protected. Hence, an adversary cannot modify or inject fake messages pretending to originate from one legitimate user.

8.6.3 IPFS Delegated Erasure Protocol

Let $h : \{0,1\}^* \rightarrow \{0,1\}^k$ be a secure hash function, and two keyed permutation $E_x : \{0,1\}^k \rightarrow \{0,1\}^n$, $e_y : \{0,1\}^k \rightarrow \{0,1\}^\nu$ for keys $x \in \{0,1\}^\lambda$ and $y \in \{0,1\}^n$, respectively, for security parameters $k, n, \nu, \lambda \in \mathbb{N}$.

The protocol is a set of five polynomial-time algorithms, namely $Keygen$, $ConKeygen$, $GenProof$, $GenDelReq$ and $CheckProof$ and is composed of three phases: initialization, content dissemination, and erasure request and verification.

- $Keygen(1^\lambda) \rightarrow mk$: A probabilistic algorithm for generating a *personal master key* $mk \in \{0,1\}^\lambda$ which is kept secret by each user.
- $ConKeygen(mk, c) \rightarrow k$: An algorithm for generating a content-based key k for the master key mk of the user that wants to submit her content $c \in \{0,1\}^*$. The generated key k can be shared with the users that should be granted content erasure. In this work, we set $ConKeygen(mk, c) = E_{mk}(h(c))$.
- $GenProof(k, c) \rightarrow s$: An algorithm for generating a proof of ownership s for a content c using a content-based key k. We instantiate $GenProof$ as: $GenProof(k, c) = e_k(h(c))$.
- $GenDelReq(k, c) \rightarrow (h, k)$: An algorithm for generating a request for erasure of content c from the user that disseminated it. Takes as input the content c and the content based key k. It outputs the erasure request which consists of the hash h of the content to facilitate its discovery and the corresponding key k that proves the ownership of the content. We instantiate this algorithm as: $GenDelReq(k, c) = (h(c), k)$.
- $CheckProof(h, k) \rightarrow \{"success", "fail"\}$ an algorithm executed by the recipient of an erasure request to determine whether she has pinned locally a content c with hash h. If this is the case, the recipient checks whether the corresponding proof s was generated using key k, which is done by simply verifying that for the hosted tuple (c, s) it holds that $s = e_k(h(c))$. The algorithm returns $success$ if the check was successful and $fail$ otherwise.

During the initialization phase, each user executes $Keygen$ to generate her personal master key $mk \in \{0,1\}^\lambda$, which is kept secret. Next, the content dissemination phase takes place, in which Alice wants to submit her content $c \in \{0,1\}^*$ to the IPFS network. Further to than simply storing and sharing c, Alice executes $ConKeygen$ and $GenProof$ to create the proof of ownership for her content. More precisely, $ConKeygen$ is realized by computing key $k = E_{mk}(h(c))$, which is subject to her personal master key and the content she wants to share. To generate her ownership proof, she uses $GenProof$ to compute $s = e_k(h(c))$. Finally, she commits the tuple (c, s) that is disseminated to the IPFS network by using the IPFS distributed sloppy hash table (DSHT) and BitSwap protocol [7]. The IPFS DSHT stores a key-value set which is spread over the participating nodes to enhance performance and scalability, by "announcing" the content stored in the nodes and which is referred to the corresponding hash $h(c)$).

Algorithm 1: Handling of an erasure request from a node.

1: On receiving a content erasure request $d = (h(c), k)$
2: **if** Content c with $h(c) = h$ is stored in the node **then**
3:     **if** $CheckProof(h, k) == True$ **then**
4:         Delete c from local storage.
5:         Forward d to neighbor nodes using DSHT ℓ times every T seconds.
6:     **end if**
7: **end if**

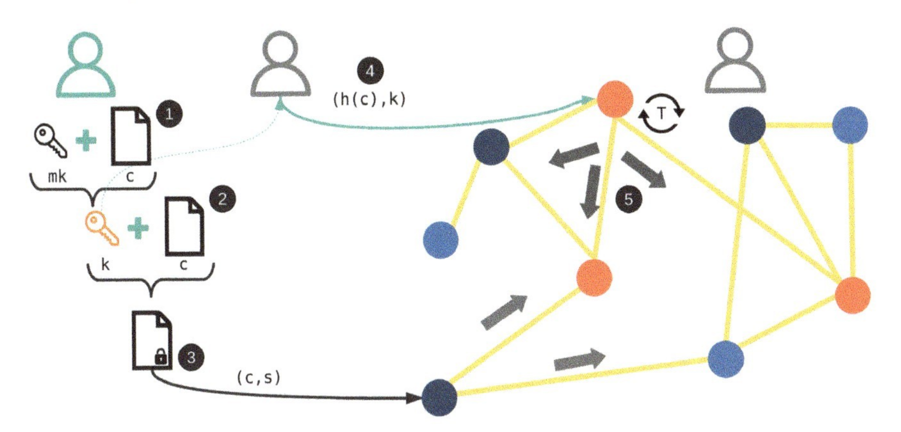

Fig. 8.1 An overview of the proposed protocol. (1) Alice uses her master key mk to create the content-dependent key k for content c. (2) Alice uses k and c to derive the proof of ownership s. (3) Tuple (c, s) is pushed to the IPFS network. (4) Alice or one of her delegates use the content-dependent key k to issue a request for erasure for c by computing $(h(c), k)$. (5) The request is disseminated to the IPFS network, and each node checks the validity of the request

In the third phase of erasure request and verification, Alice (or one of her delegates) decides to erase content c from IPFS. Hence, she uses the $GenDelReq$ and sends her request as a tuple $d = (h(c), k)$ to the network. Any receiving node may now use $CheckProof$ to first locate the content using $h(c)$, and then to verify whether the erasure request is valid, i.e. $s = e_k(h(c))$. Finally, to minimize the network overhead, the receiving node forwards the erasure request d to all its neighbours holding the file by using DSHT and the file's corresponding hash $h(c)$. To be noted that this action is already possible in IPFS with functions like (*ipfs dht findprovs* $<h(c)>$). Hence, the required modification would be made by simply changing the $<h(c)>$ value to $(<h(c)>, k)$.

Figure 8.1 illustrates a workflow overview of the proposed protocol starting from the creation of mk, up to the point where other nodes check the validity of the erasure request sent by a delegate for a given content c. Finally, Algorithm 1 outlines the handling process of an erasure request from a node. Note that the process is repeated ℓ times every T seconds to accommodate for offline nodes. Both of these values are constants and can be set either by each node individually or by computing specific parameters about the node participation distribution and the probability of storing a content.

8.6.4 Security Proof

Next, we formally prove the properties of the proposed protocol.

Theorem 8.1 *Alice's personal master key is secure against any PPT adversary if the keyed permutation E is secure.*

Proof For the sake of brevity and convenience, we prove the theorem for the worst-case scenario, which is the case of a malicious delegate. Contrary to any other adversaries, a delegate for erasing any of Alice's contents is the only one who has some output directly linked to Alice's personal master key.

Let us assume that a malicious delegate of Alice, Malory (from now on denoted as $\mathbb{M}$) wants to extract the personal master key of Alice X. Therefore, we assume that $\mathbb{M}$ has access to a set of $m > 0$ delegated keys $K_{C_j X} = E_X(h(c_j)), j \in \{1, 2, \ldots, m\}$ for the corresponding contents $c_j, j \in \{1, 2, \ldots, m\}$. To extract X, $\mathbb{M}$ must perform a known-plaintext attack to E. Since E is a secure keyed permutation this is not possible, so X is secure from $\mathbb{M}$.

Theorem 8.2 *A PPT adversary cannot forge an erasure request for any given tuple* (c, s) *if the keyed permutation e is secure.*

Proof Let us assume that a PPT adversary $\mathbb{M}$ wants to forge an erasure request for a given tuple (c, s). $\mathbb{M}$ needs to find $\kappa \in \{0, 1\}^k$ such that $e_\kappa(h(c)) = s$. Since e is a secure keyed permutation, κ cannot be computed in probabilistic polynomial time. Therefore, an erasure request for any tuple (c, s) cannot be forged by $\mathbb{M}$.

Theorem 8.3 *Given two tuples of the form* $(c, s) = \big(c, e_{K_{CX}}(h(c))\big)$, $(c', s') = \big(c', e_{K_{C'X'}}(h(c'))\big)$, *a PPT adversary cannot determine whether they belong to the same user if the keyed permutations e and E are secure.*

Proof To determine whether two tuples (c, s) and (c', s') belong to the same user, one has to determine whether $X = X'$. Since the keyed permutations e and E are secure, one cannot recover either K_{CX} nor $K_{C'X'}$ and from them X and X' to determine whether they are equal or not. Therefore, a PPT adversary cannot determine whether two tuples (c, s) and (c', s') belong to the same user.

Corollary 8.1 *The proposed erasure protocol is anonymous.*

Proof The proof of the theorem above follows from the anonymity that IPFS provides and the proof of Theorem 8.3.

8.6.5 Protocol Efficiency

The proposed protocol has a minimal footprint as the additional overhead for the proofs-of-ownership in terms of storage space is equal to the length of the encryption of a hash. In terms of creating a proof, one has to perform two encryptions with a symmetric cipher, and two hashes. Similarly, for validating the request, which requires checking both whether a given content exists and whether an erasure request is valid, one hash and one decryption with a symmetric cipherone has to be performed.

To validate the efficacy of our proposal, we implemented the proposed protocol proofs in Python 2.7. We used AES in CBC mode with 256-bit keys and SHA-256 hash function. Without using parallelization, we generated 1000 random files

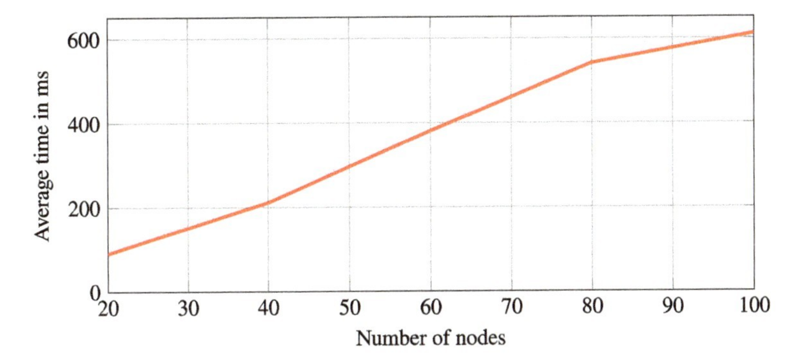

Fig. 8.2 Performance analysis of the simulated P2P network with different number of participating nodes

of 1MB on a system running with an Intel Core i7-6700K CPU at 4.00GHz and 16GB of RAM on Ubuntu 19.04. We then created a master key for the user, and for each file we computed the corresponding keys, the proof-of-ownership and the verification time. The average time for generating the content-dependent keys, the proofs-of-ownership, and the verification of the proof are 2.011 ms, 2.023 ms, and 1.99 ms, respectively, which can be considered minimal. In regard to the rest of functionalities, we consider only existing functions of the IPFS (see Sect. 8.6.3), so that minimal modifications are required.

To further study the performance of our proposed protocol we used the NS-3 network simulator[4] to simulate a p2p communication that spreads data packets among the network nodes [8]. As shown in Fig. 8.2, in our simulation we designed a node that disseminates the erasure request to different numbers of participating nodes ranging between 20 and 100 nodes. The size of the packet travelling through the network is typically 64 bytes. The X-axis represents the number of participating nodes, which had been increased statically to study the time impact when the network expands. The Y-axis represents the average time required in each case (i.e., the time required to disseminate the request) with no deviation of the time. In terms of scalability, the time increases as more nodes are added to the network; hence a linear relationship exists with a positive line slope. Extending our previous experiment, we studied the impact of the packet size on the IPFS network time latency to analyse the required time for handling the erasure request. As shown in Fig. 8.3, we validated our network for cases where the packet size (i.e. user key and the hash of the file) changes, ensuring the scalability of the method and its resilience to future variations of the hash size.

[4] http://www.nsnam.org/.

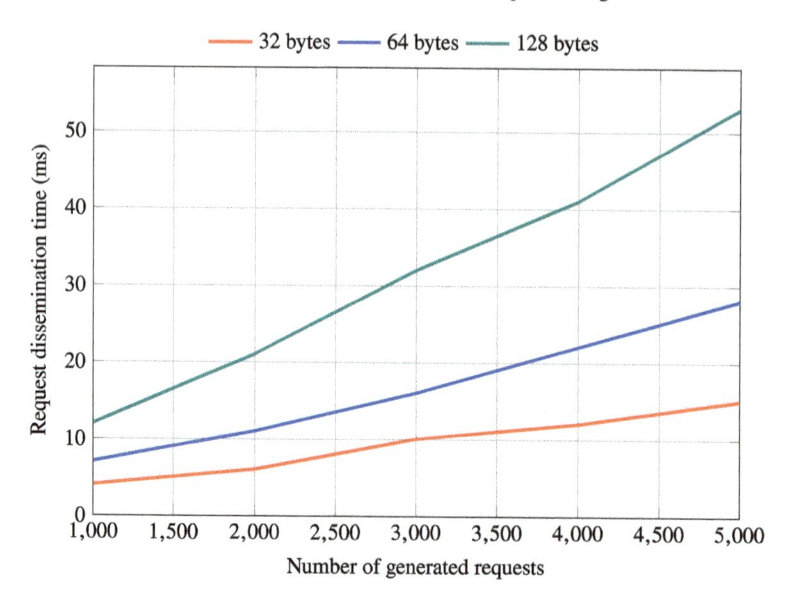

Fig. 8.3 Simulation of disseminating large-scale nodes communication with different packet sizes in the IPFS network to validate a request

8.6.6 *Limitations and Countermeasures*

Despite the several afore-mentioned benefits provided by our proposed erasure mechanism, its integration into the IPFS protocol presents also some limitations that should be pointed out. One limitation arises due to the duplication attack according to which a user could claim the ownership of a content that was previously deleted by committing it again with her own key. A simple way to prevent this from happening is to store in a parallel structure (e.g. a Merkle DAG) the erasure requests made by the users. For this purpose, the protocol could be further expanded to check if a file that is going to be added in the IPFS already existed in the system. On top of this, the IPFS protocol could be further updated to prevent users from adding an existing file. This can be achieved either by forcing the execution of the following commands:

```
ipfs refs local | grep <hash>
```

and

```
ipfs dht findprovs <key>
```

or by directly checking if the file exists through one of the gateways. Obviously, these checks have an implicit overhead of several milliseconds [6, 9] and therefore, the trade-off between performance and security must be carefully examined and discussed prior to their implementations. Aside from these checkings, to enhance further the scalability and performance, the erasure requests could have some specific time to live (TTL). Nonetheless, it is worth noticing that such kind of attacks can be performed in any kind of platform or database and hence establishing a mitigation strategy against these attacks is quite complex. Yet, our introduced approach partially avoids this kind of malicious behaviour.

Another issue, closely related to the previous one, is the case of a malicious user which either uploads content from competitors, or that is legally owned by other authors, or any other kind of copyright-protected files, to dishonestly claim their ownership. Similarly to the duplication attack, this case is not specific to our proposed protocol nor to the IPFS since it may occur in all available storage systems. A viable workaround to this issue would be the implementation of an ownership claim protocol which involves an identity management system supported by a consensus mechanism. Nevertheless, while this implementation could indeed provide a nice-to-have feature, we argue that it is beyond the essence of the IPFS and as such, it is out of the scope of this book.

References

1. E. Politou, F. Casino, E. Alepis, C. Patsakis, Blockchain mutability: challenges and proposed solutions. IEEE Trans. Emerging Top. Comput. 1–1 (2019) https://doi.org/10.1109/TETC.2019.2949510
2. M. Scherer, *Performance and Scalability of Blockchain Networks and Smart Contracts* (Umea University, Sweden, 2017)
3. European Union, Regulation (EU) 2016/679 of the European Parliament and of the Council of 27 April 2016 on the protection of natural persons with regard to the processing of personal data and on the free movement of such data, and repealing Directive 95/46/EC (General Data Protection Regulation). Off. J. Eur. Union **L119** 1–88
4. E. Politou, E. Alepis, C. Patsakis, Forgetting personal data and revoking consent under the GDPR: challenges and proposed solutions. J. Cybersecu. **4**(1), tyy001 (2018)
5. F. Casino, E. Politou, E. Alepis, C. Patsakis, Immutability and decentralized storage: an analysis of emerging threats. IEEE Access **8**, 4737–4744 (2020)
6. C. Patsakis, F. Casino, Hydras and IPFS: a decentralised playground for malware. Int. J. Inf. Secur. (2019)
7. J. Benet, *IPFS-Content Addressed, Versioned, p2p File System* (2014). arXiv:14073561
8. L. Campanile, M. Gribaudo, M. Iacono, F. Marulli, M. Mastroianni, Computer network simulation with ns-3: A systematic literature review. Electronics **9**(2), 272 (2020)
9. B. Confais, A. Lebre, B. Parrein, Performance analysis of object store systems in a fog/edge computing infrastructures, in *2016 IEEE International Conference on Cloud Computing Technology and Science (CloudCom)*, pp. 294–301 (2016). https://doi.org/10.1109/CloudCom.2016.0055

Chapter 9
Privacy in the COVID-19 Era

Abstract The sudden outbreak of COVID-19 at the late 2019 has brought enormous hurdles globally to our everyday lives and to our society. In order to mitigate the impact of the pandemic and to control the dissemination of the coronavirus, governments worldwide have taken extreme surveillance measures which most of the times invade to individuals' privacy and contradict data protection principles. In this chapter, we investigate the interplay between privacy and COVID-19 pandemic and we explore the invasive digital technologies such as the mobile contact tracing apps and health immunity passports that have been utilized to monitor people's lives.

9.1 Introduction

In 2019, a new form of coronavirus made an outbreak in Wuhan (China) and soon spread throughout the globe, creating a global pandemic. To counter the spread of the virus many countries resorted to the use of lockdowns and constraining measures regarding people's free movement. All of a sudden, our life changed dramatically to mitigate the fast spread of COVID-19 and our everyday reality became less carefree: the financial difficulties due to the suspension of businesses' operations, the psychological effects of quarantine and social distancing, and even worst the loss of the beloved ones, are some critical consequences that most people have been dealing with since the beginning of the pandemic. Unfortunately, however, as the virus seems not to be easily confined, citizens and businesses have to radically realign their lives to this new reality.

Given that the pandemic has already costed numerous human lives and the hospital capacities are being continuously heavily tested, the situation calls for drastic measures. Globally, states and public health organizations have been relying greatly on digital technologies and big data analytics to implement measures against the prevalence of the new virus. To this end, massive amount of personal health information is collected by public and private entities to track citizens' movements and to facilitate traveling or other social activities, eliminating thereby the risk of coronavirus contamination. For instance, biobank registers containing information about

E. Politou et al., *Privacy and Data Protection Challenges in the Distributed Era*,
Learning and Analytics in Intelligent Systems 26,
https://doi.org/10.1007/978-3-030-85443-0_9

the infected population, databases about vaccinated individuals, as well as digital registers of the overall health status of individuals, have been developed by many affected countries.

By far, the most prevailing technological measures adopted by many countries to control the pandemic are the mobile tracing applications utilizing fine-grained personal big data to monitor user's contacts, daily routines, and health statuses, as well as the health immunity passports/certificates. Mobile applications, in particular, play a central role in the fight against the spread of the virus since they can successfully meet the needs for rapid contact monitoring. Various mobile solutions that detect holder's contacts have been emerged to trace the possibly infected individuals and to isolate them. Countries such as Israel are using counter-terrorism cyber technologies to fight the uncontrolled dissemination of the pandemic,[1] whereas others like China and Russia leverage advanced face recognition technology to police their lockdowns and to track possible offenders.[2,3,4] While Asian countries such as Taiwan, Singapore, and South Korea are considered among the most highly tech nations which lead the extreme tracking of their citizens,[5,6,7,8] their European counterparts are closely following in this race for urgent pandemic prevention.[9,10] Hence, the European Parliament in its resolution in May 2020 underlined that any digital measures against the pandemic must be in full compliance with data protection and privacy legislation.[11] Among the basic notions that have been stressed by the Parliament was the non-obligatory use of these mobile apps and their strict data retention clauses to limit their use only during the pandemic period.

Nevertheless, the security and privacy challenges arising from these technologies are tremendous. To this end, below we first introduce the basic building blocks and technical charasteristics of the contact tracing mobile applications and health immunity passports, and then we discuss the impact of these digital surveillance efforts on data protection and individuals' privacy.

---

[1] https://www.reuters.com/article/us-health-coronavirus-israel/israel-to-use-anti-terror-tech-to-counter-coronavirus-invisible-enemy-idUSKBN21113V.

[2] https://www.bbc.com/news/av/world-europe-52157131.

[3] https://abcnews.go.com/International/russia-facial-recognition-police-coronavirus-lockdown/story?id=70299736.

[4] https://www.reuters.com/article/us-health-coronavirus-facial-recognition-idUSKBN20W0WL.

[5] https://www.bloomberg.com/opinion/articles/2020-04-22/taiwan-offers-the-best-model-for-coronavirus-data-tracking.

[6] https://hbr.org/2020/04/how-digital-contact-tracing-slowed-covid-19-in-east-asia.

[7] https://asia.nikkei.com/Spotlight/Coronavirus/Singapore-mandates-use-of-tracing-app-as-it-fights-COVID-s-spread.

[8] https://www.bbc.com/news/technology-52681464.

[9] https://ec.europa.eu/info/live-work-travel-eu/coronavirus-response/travel-during-coronavirus-pandemic/mobile-contact-tracing-apps-eu-member-states_en.

[10] https://www.mobihealthnews.com/news/emea/covid-19-alert-and-warning-apps-protect-lives-and-livelihoods-says-european-commission.

[11] https://www.europarl.europa.eu/doceo/document/TA-9-2020-0054_EN.pdf.

9.2 Contact Tracing Apps

Traditionally, contact tracing is performed through specialized interviews of the individuals by the authorities and it aims at identifying the people with which an individual has contacted in a specific timeframe. However, things are more complicated for health-related issues, as people do not necessarily should have talked with others to consider them a contact. The co-location of two individuals in a specific area for enough duration is enough to infect each other. Therefore, while people could trace their contacts with high accuracy in the family and work environment, it is almost impossible to do the same in the context of public transportation, shopping, or even recreation time etc.

To address this critical problem, given the global scale and timing constraints, technological means were considered the most eligible solutions. Due to the continuous use of smartphones from billions of users, researchers and governments have proposed the introduction of mobile apps which can exploit their integrated sensors to allow for the seamless and continuous contact tracing. Indeed, as suggested by Ferreti et al. [1], fast contact tracing can facilitate in keeping the reproduction number R of COVID-19 at low values. The solutions that are proposed in the current literature vary on the employed infrastructure (centralized, decentralized, and hybrid). Nevertheless, regardless of which infrastructure is used, the vast majority of the apps use Bluetooth to exchange or just to broadcast a contact beacon. The choice of Bluetooth to transmit contact beacons was made for the following reasons:

- The range of Bluetooth is bounded to approximately 10m which can be efficiently cover the range within which that healthcare practitioners consider that two individuals have contacted (1–2 m).
- Bluetooth does not require prior network connectivity parameterization or pairing procedure between unknown devices.
- Bluetooth has low power consumption.

Obviously, keeping track of which people someone has contacted raises far too many privacy issues as it can potentially disclose the social contacts of people, their location, their religious and sexual preferences, etc.

In the centralized setting, a service provider collects the user data through a registration phase and then assigns them a temporary ID which is periodically updated; usually in the scale of minutes. This temporary ID is broadcasted by each device along with the energy needed to transmit the signal. While the latter is used to compute the proximity of the user, the former is used to prevent recording the contact multiple times. This information is recorded in each device along with a timestamp and is sent to the central server at regular predefined intervals. Once a user is diagnosed with COVID-19, the information is sent to server and the tracing begins. The central server can map the temporary IDs back to the original users and, based on their proximity, issue an alert that they have been in contact with an infected person.

However, in the decentralized setting there is not any registration phase and the users periodically generate their temporary IDs on their own. Once they meet someone, the devices exchange a token and then both tokens are sent to a server or a distributed architecture. Should a user be diagnosed positive to COVID-19, her device submits the tokens of her recent contacts. Clearly, the server or the distributed architecture serves as an anonymous bulletin board. Users may download locally these tokens and determine whether they came in contact with someone diagnosed with COVID-19.

In the hybrid model, the users register to a service provider to create ephimeral IDs which are periodically updated and used to initiate a Diffie-Hellman key exchange from which two tokens are generated. These tokens can then be used for tracing whether an individual has come in contact with a COVID-19 patient. Note that the exchange of these tokens is performed through proxies whereas only health authorities can update the status of each user. Therefore, the central server does not have access to any sensitive information. For more information on contact tracing applications, the interested reader may refer to [2, 3].

Beyond any doubt, the centralized scenario implies serious privacy issues for the users. On the other hand, the decentralized scenario can be exploited to introduce high amounts of false positives. In the hybrid scenario, the user exposure highly depends on the implementation. Taking into account the complexity of the problem, the EDPB issued some generic guidelines on the use of location data and contact tracing tools to prevent potential exploitation of sensitive data of EU citizens.[12] To this end, contact tracing should be voluntarily without forcing consequences to those who do not adopt it, and it should use only proximity information and not location. User identification should not be possible from the data that are exchanged from application and only the necessary information must be sent from the device to the outside world. Given their market share and the need for such solutions, Apple and Google partnered and jointly developed the *Exposure Notifications*,[13] which matches the hybrid scenario discussed above.

Although the above efforts aim at facilitating the isolation of the infected subjects and intercepting the pandemic, their effectiveness has been questioned [4–6] since, for instance, there is a high false positive rate which results to far too many users undergo a COVID-19 testing, creating an additional bottleneck in the healthcare sector. The way these apps have been used throughout the globe differs. For instance, Chinese apps have been reported to use massive data from users[14] and automate the classification of users based on risk exposure and using QR codes to allow access to services. The *Exposure Notifications* protocol introduces potential security issues, as recently shown by Gvili [7]. Even more, it has been shown that the Google's implementation of this protocol leaked location data.[15] Additionally, Kouliaridis et al. [8] and Hatamian et al. [9], individually, have found several issues in Android contact tracing apps. Sun et al. [10] have recently introduced COVIDGUARDIAN

[12] https://edpb.europa.eu/our-work-tools/our-documents/guidelines/guidelines-042020-use-location-data-and-contact-tracing_en.

[13] https://www.google.com/covid19/exposurenotifications/.

[14] https://www.nytimes.com/2020/03/01/business/china-coronavirus-surveillance.html.

[15] https://www.zdnet.com/article/contact-tracing-apps-android-phones-were-leaking-sensitive-data-find-researchers/.

to automatically analyse COVID-19 apps for security issues with rather alarming results. Last but not least, there have been numerous reports of such apps leaking personal data online.[16]

Despite the inherent vulnerabilities of mobile apps, one should also consider that these apps highly depend on Bluetooth technology which implies further privacy exposure. For instance, there are several attacks targeting Bluetooth enabled devices [11–14] which can be used to exfiltrate user sensitive data or inject false information in the system. Note that while the Bluetooth may transmit a random-looking beacon, driver issues may disclose the identity of the device and facilitate the tracking of individuals beyond the scope of COVID-19 apps [15, 16].

9.3 Immunity Passports

Going a step further from tracing apps and as a means to create a safe transition to the post COVID-19 era, governments are currently introducing the concept of *Immunity passports*. In essence, to allow people to travel across countries and to attend events where big crowds are expected, governments are trying to push the use of certificates which prove that a certain individual has been fully vaccinated, or has recently recovered from COVID-19, or has been tested negative during the past few days.

Their introduction, however, has initiated a lot of skepticism and debates regarding their ethics and legality [17–20]. The raised issues stem from the arising privacy concerns, the potential marginalization of social groups, and the possible discrimination of individuals. Despite these hesitations, governments have proceeded to the implementation of such measures. The EU has introduced the *EU Digital COVID Certificate*[17] to facilitate safe free movement of citizens across the EU. The key features of this certificate are that it is free of charge in both digital and paper format, it is valid for all EU countries and QR codes are used to validate it. Nevertheless, a proof of identity is needed for one to verify certificate's validity. The potential misuse of such certificate when not used for traveling but for accessing facilities and services where one would not had to use a form of identity is triggering more debates. Potentially, in the near future, an anonymized certificate with only the QR and the photo of its holder will facilitate its integration and its wide adoption.

Academic literature provide possible solutions the above problems by following a decentralized approach which is heavily dependent on blockchains [21–25]. The main issue that blockchains can resolve is to provide auditability in a multi-stakeholder scenario where the users are not trusting each other and no fully trusted party exists. Therefore, blockchains serve as a means to keep an audit trail of the tests and/or

[16] https://www.reuters.com/article/us-health-coronavirus-netherlands-datapr-idUSKBN29Y1H3, https://www.bbc.com/news/technology-52725810.

[17] https://ec.europa.eu/info/live-work-travel-eu/coronavirus-response/safe-covid-19-vaccines-europeans/eu-digital-covid-certificate_en.

vaccinations of individuals where citizens may selectively allow others to access their trace. Thus, access and key management, as well as identity management are becoming a huge issue. Especially for identity management, many researchers opt for self sovereign identities to allow more freedom to the users, while others prefer stronger authentication using, e.g. FIDO.

9.4 Privacy and Data Protection in the Pandemic

While nowadays privacy is an important priority - both for individuals and institutions - safeguarding personal data in the COVID-19 era is becoming an even more challenging task than before this pandemic. Balancing privacy and public health interests is definitely not easy given this unexplored situation we are living in. In this regard, the Council of Europe has announced that *"while it is crucial to make clear that data protection can in no way be an obstacle to save human lives, it is equally crucial to reaffirm that the exercise of human rights and notably the rights to privacy and to data protection are still applicable.*[18] In the same context, the EDPB has published a statement on the processing of personal data in the context of the COVID-19 outbreak explaining that data protection rules and regulations such as the GDPR do not hinder measures taken in the fight against the coronavirus pandemic.[19] They also underlined that, even in these exceptional times, data controllers must ensure the protection of the personal data of the data subjects.

Yet, even though the EU Member States may have taken privacy concerns more seriously - given the strict provisions foreseen in the GDPR—when developing these tracking mobile apps, public authorities worldwide have loosened their approach to privacy in view of the urgent health measures needed to constrain the coronavirus contagion. Nevertheless, this immense collection, process and dissemination of personal sensitive information has put privacy into the spotlight and has alarmed privacy advocates and international data protection authorities who argue that whilst the measures described above may be potentially justifiable on public health grounds, the sharing of such sensitive information should be done in a privacy protective manner. Hence, it is not surprising that this new surveillance era is being closely monitored by international privacy specialists and human rights organizations. Several maps[20,21,22] illustrating the type and the range of COVID-19 mobile surveillance per country have

---

[18] https://www.coe.int/en/web/data-protection/covid-19-data-protection.

[19] https://edpb.europa.eu/sites/default/files/files/file1/edpb_statement_2020_processingpersonal dataandcovid-19_en.pdf.

[20] https://www.coe.int/en/web/data-protection/contact-tracing-apps.

[21] https://onezero.medium.com/the-pandemic-is-a-trojan-horse-for-surveillance-programs-around-the-world-887fa6f12ec9.

[22] https://www.nortonrosefulbright.com/en-lu/knowledge/publications/d7a9a296/contact-tracing-apps-a-new-world-for-data-privacy.

been developed to demonstrate the degree to which privacy rights have been shifted to protect the ultimate good of public health.

Various data protection agencies have also published guidelines on how data controllers should respect privacy principles while ensuring that public health measures are appropriately implemented to protect the citizens' health. The Global Privacy Assembly (GPA) Executive Committee has issued a joint statement on the importance of privacy by design in the sharing of health data for domestic or international travel requirements during the COVID-19 pandemic.[23] The GPA Executive Committee recalls that while big data and technology can be important tools to help fight the COVID-19 pandemic, they have intrinsic limitations and can merely leverage the effectiveness of other public health measures and need to be part of a comprehensive public health strategy to fight the pandemic. The principles of effectiveness, necessity, and proportionality must guide any measure adopted by government and authorities that involve processing of personal data to fight COVID-19. Furthermore, data protection authorities worldwide have already issued guidance providing information and frequently asked questions pertaining to data processing and COVID-19.[24]

According to the guidelines from the various European DPAs, public institutions have the possibility to rely on the legal basis from Article 9 (2) of the GDPR which allows the processing of health data when the "*processing is necessary for reasons of public interest in the area of public health, such as protecting against serious cross-border threats to health*". Moreover, the principles of lawfulness, fairness, transparency, proportionality, purpose limitation, data minimisation, accuracy, storage limitation, integrity and confidentiality need to be taken into account when processing personal data under the conditions of pandemic. Yet, even though these principles are endorsed by all data protection guidelines and the GDPR itself, the outbreak of the COVID-19 virus is the first real situation in which they need to be tested when the fundamental rights to health and, by extent, to life, are at stake.

In view of the above, the EDPS underlined that data protection rules currently in force in Europe are flexible enough to allow for various measures taken in the fight against pandemics.[25] To support their argument, the EDPS declared that the approach to handle the emergency in the most efficient, effective and compliant way possible would be the elements for data anonymisation, the implementation of equally effective security measures that are bound by strict confidentiality obligations and prohibitions on further use of the personal data, and the limited retention of these data.

However, given that balancing fundamental rights is not an easy neither a straightforward case , the GDPR currently faces its first real obstacle since its enforcement in 2018. As it has been pointed out, this global health crisis is actually the first real case scenario in which the GDPR needs to be tested to prove its efficiency in practice when

[23] https://globalprivacyassembly.org/gpa-executive-committee-joint-statement-on-the-use-of-health-data-for-domestic-or-international-travel-purposes/.

[24] https://iapp.org/resources/article/dpa-guidance-on-covid-19/.

[25] https://edps.europa.eu/sites/edp/files/publication/20-03-25_edps_comments_concerning_covid-19_monitoring_of_spread_en.pdf.

personal data are exposed to such a great scale. Undeniably though, privacy experts argue that this can be seen also as a great opportunity for the GDPR to demonstrate its flexibility when matters of public interest are in jeopardy.

9.5 Conclusions

Unfortunately, at the time of writing, despite the steep decline in the global number of COVID-19 patients and the wide use of several vaccines, the situation cannot be considered safe. Indeed, mutations of the virus make experts think that we have to learn to live with COVID-19. Therefore, many of the privacy issues that were reported in this chapter are expected to continue to be relevant in the near and long future. While general lockdowns may be eventually prevented, the use of contact tracing apps, digital health certificates as well as massive remote working are expected to create new norms.

Regardless of its necessity, the continuous monitoring of individuals and the massive collection of their personal data is not easily tolerated, especially in western societies. Consequently, the conflicts between public health interests and data protection principles are more relevant today than any other time. Following the enforcement of GDPR in 2018, a new more intense debate on balancing what data are collected, by whom, who processes it, and when, is anticipated. The latter is expected not only due to its necessity for fighting the COVID-19 virus or any other future pandemic, but because of the spike in cybercrime that was noticed during the COVID-19 pandemic[26] and the debate that has started for *Security through encryption and security despite encryption*[27] within the EU.

References

1. L. Ferretti, C. Wymant, M. Kendall, L. Zhao, A. Nurtay, L. Abeler-Dörner, M. Parker, D. Bonsall, C. Fraser, Quantifying SARS-COV-2 transmission suggests epidemic control with digital contact tracing. Science **368**(6491) (2020)
2. N. Ahmed, R.A. Michelin, W. Xue, S. Ruj, R. Malaney, S.S. Kanhere, A. Seneviratne, W. Hu, H. Janicke, S.K. Jha, A survey of covid-19 contact tracing apps. IEEE Access **8**, 134577–134601 (2020)
3. T. Martin, G. Karopoulos, J.L. Hernández-Ramos, G. Kambourakis, I. Nai Fovino, Demystifying covid-19 digital contact tracing: a survey on frameworks and mobile apps. Wirel. Commun. Mobile Comput. (2020)
4. J. Almagor, S. Picascia, Exploring the effectiveness of a covid-19 contact tracing app using an agent-based model. Sci. Rep. **10**(1), 1–11 (2020)
5. L. Maccari, V. Cagno, Do we need a contact tracing app? Comput. Commun. **166**, 9–18 (2021)

[26] https://www.europol.europa.eu/newsroom/news/covid-19-sparks-upward-trend-in-cybercrime.

[27] https://data.consilium.europa.eu/doc/document/ST-13084-2020-REV-1/en/pdf.

6. L. White, P. Van Basshuysen, Without a trace: Why did corona apps fail? J. Med. Ethics (2021)
7. Y. Gvili, Security analysis of the covid-19 contact tracing specifications by Apple inc. and Google inc. IACR Cryptol ePrint Arch **2020**, 428 (2020)
8. V. Kouliaridis, G. Kambourakis, E. Chatzoglou, D. Geneiatakis, H. Wang, Dissecting contact tracing apps in the android platform. Plos one **16**(5), e0251867 (2021)
9. M. Hatamian, S. Wairimu, N. Momen, L. Fritsch, A privacy and security analysis of early-deployed covid-19 contact tracing Android apps. Empir. Softw. Engi. **26**(3), 1–51 (2021)
10. R. Sun, W. Wang, M. Xue, G. Tyson, S. Camtepe, D.C. Ranasinghe, An empirical assessment of global covid-19 contact tracing applications, in *2021 IEEE/ACM 43rd International Conference on Software Engineering (ICSE)* (IEEE, 2021), pp. 1085–1097
11. D. Antonioli, N.O. Tippenhauer, K. Rasmussen, Bias: bluetooth impersonation attacks, in *2020 IEEE Symposium on Security and Privacy (SP)* (IEEE, 2020), pp. 549–562
12. K. Haataja, K. Hyppönen, S. Pasanen, P. Toivanen, Bluetooth Security Attacks: Comparative Analysis, Attacks, and Countermeasures (Springer, 2013)
13. D.Z. Sun, Y. Mu, W. Susilo, Man-in-the-middle attacks on secure simple pairing in bluetooth standard v5. 0 and its countermeasure. Pers. Ubiquitous Comput. **22**(1), 55–67 (2018)
14. Y. Zhang, J. Weng, R. Dey, Y. Jin, Z. Lin, X. Fu, Breaking secure pairing of bluetooth low energy using downgrade attacks, in *29th {USENIX} Security Symposium ({USENIX} Security 20)* (2020), pp. 37–54
15. J.K. Becker, D. Li, D. Starobinski, Tracking anonymized bluetooth devices. Proc. Priv. Enhancing Technol. **3**, 50–65 (2019)
16. M. Cominelli, F. Gringoli, P. Patras, M. Lind, G. Noubir, Even black cats cannot stay hidden in the dark: full-band de-anonymization of bluetooth classic devices, in *2020 IEEE Symposium on Security and Privacy (SP)* (IEEE, 2020), pp. 534–548
17. R.C. Brown, D. Kelly, D. Wilkinson, J. Savulescu, The scientific and ethical feasibility of immunity passports. Lancet Infect. Diseases (2020a)
18. R.C. Brown, J. Savulescu, B. Williams, D. Wilkinson, Passport to freedom? immunity passports for covid-19. J. Med. Ethics **46**(10), 652–659 (2020)
19. T. Osama, M. Razai, A. Majeed, Covid-19 vaccine passports: access, equity, and ethics. BMJ **373** (2021)
20. A.L. Phelan, Covid-19 immunity passports and vaccination certificates: scientific, equitable, and legal challenges. Lancet **395**(10237), 1595–1598 (2020)
21. A. Bansal, C. Garg, R.P. Padappayil, Optimizing the implementation of covid-19 "immunity certificates" using blockchain. J. Med. Syst. **44**(9), 1–2 (2020)
22. H. Choudhury, B. Goswami, S.K. Gurung, Covidchain: an anonymity preserving blockchain based framework for protection against Covid-19. Glob. Perspect. Inf. Secur. J. 1–24 (2021)
23. H.R. Hasan, K. Salah, R. Jayaraman, J. Arshad, I. Yaqoob, M. Omar, S. Ellahham, Blockchain-based solution for covid-19 digital medical passports and immunity certificates. IEEE Access (2020)
24. J.C. Polley, I. Politis, C. Xenakis, A. Master, M. Kępkowski, On an innovative architecture for digital immunity passports and vaccination certificates (2021). arXiv:210304142
25. L. Ricci, D.D.F. Maesa, A. Favenza, E. Ferro, Blockchains for covid-19 contact tracing and vaccine support: a systematic review. IEEE Access **9**, 37936–37950 (2021)

Chapter 10
Open Questions and Future Directions

Abstract Modern technological advancements such as mobile ubiquitous computing and decentralized p2p networks rely on the collection, processing and sharing of vast amount of personal information which—when combined with big data and machine learning techniques—pose significant challenges to the rights of privacy and data protection. The GDPR, seeking to regulate the uncontrollable dissemination of personal data across the Internet and to give control of their data back to individuals, enshrines the RtbF to allow for the retroactive and permanent erasure of their personal data from all the places to which they have been disseminated. However, the issues arising from the GDPR data protection provisions, and in particular by the RtbF, are not trivial since they overlap with other rights such as the right to information and to free speech, with organizational practices such as the backup and archiving procedures, and with novel state-of-the-art technologies such as algorithmic profiling, ubiquitous computing, and decentralized p2p architectures. Following our analyses in the preceding Chapters, hereafter we identify any remaining open research questions and we highlight potential future directions towards the effective alignment of state-of-the-art technologies and business practices with the privacy and data protection principles of the GDPR.

10.1 Introduction

Modern technological advancements, such as mobile ubiquitous computing and sensing as well as big data, rely more and more on the collection and processing of vast amount of personal information. At the same time, there is little doubt that over the next few years blockchain and other decentralized p2p technologies that share information across peers without a single entity controlling the network, will change the landscape of economic policy significantly. However, as Ohm outlines in [1], even though "*the benefits [of such technologies] outweigh the costs*", this is "*merely luck*" provided that "*precautions against misuse were not discussed widely*". Indeed, privacy and ethical issues arise continuously from the use of current state-of-the-art technologies, whereas traditional values such as the rights to privacy and to data protection are increasingly under dispute [2].

E. Politou et al., *Privacy and Data Protection Challenges in the Distributed Era*,
Learning and Analytics in Intelligent Systems 26,
https://doi.org/10.1007/978-3-030-85443-0_10

According to a newspaper article[1] a few years ago, privacy nowadays seems to become an obsolete notion, destined to be dropped from our vocabulary. Considering the volume of personal information collected by the current technology and the amount of inferred personal data for which people do not even know their existence, this belief is not far from the truth. On that account, some scholars argue that the regulation of the Internet excesses is necessary in order to gain the benefits of its substantial breakthroughs and prevent privacy harms, even though in doing so it may need to sacrifice, at least a little, important counter values, like innovation, free speech, and security [3, 4]. Nevertheless, free speech advocates claim that regulating the Internet is a challenging task as it is important that laws do not unduly infringe freedoms or deter innovation [5].

The GDPR, seeking to regulate the uncontrollable dissemination of personal data across the Internet and to give control of their data back to individuals, anticipates the RtbF to allow for the permanent erasure of their personal data across all the Internet places to which they have been disseminated. Yet, scholars and policymakers have been extensively clashing over the way that the rights to privacy, to data protection, to freedom of expression, and to be informed should be balanced in the online world under the GDPR regime. While in principle privacy and freedom of expression and information have equal weight in Europe, balancing privacy-related and freedom of expression/information-related interests will always remain difficult [6]. Beyond any doubt, the issues arising from the data protection provisions introduced by the GDPR, and in particular by the RtbF, are not trivial. In fact, they lie at the heart of intensively discussed and disputed areas since they are due to overlap with other rights, such as the right to information and to free speech; organizational practices, such as backup and archiving procedures; and innovative state-of-the-art technologies, such as big data algorithmic profiling, ubiquitous computing, and decentralized p2p architectures.

Following our analysis presented in the preceding Chapters, hereafter we identify open research questions and we highlight potential future directions towards the effective alignment of state-of-the-art technologies and business practices with the privacy and data protection principles of the GDPR. Above all, we believe that our discussion below can be a valuable contribution both to the academia and to the industry, and—by extension—to the well-being of individuals.

10.1.1 Forgetting Implementation Standards

Taking into account that personal data represent not only money but also power whose exercise may affect society and individuals in an unprecedented way, most experts agree that the fragile and complex balance between individual rights and collective knowledge should not be exclusively entrusted to market dynamics. Instead, regulators and public authorities should always ensure a level playing field between consumers and businesses in order to guarantee the respect of the fundamental rights

[1] https://www.nytimes.com/2014/05/21/opinion/friedman-four-words-going-bye-bye.html.

of individuals and the freedom of expression [5, 7]. With this in mind, we recognize that the legislation of the RtbF under the GDPR is a step in the right direction. Nonetheless, it should not be seen as a cure-all or a silver bullet that will apply immediately to all domains and solve all the problems [5]. Rather, we argue that business-level implementation guidelines and low-level business specifications, as well as technology-agnostic technical standards and modelling, should be put in place in a clear and cross-platform manner to cater for all pragmatic exceptions and complexities. In that regard, formal legal and technical European bodies should take appropriate measures and initiate work on that direction in the immediate future. Even then, however, any roadmaps and standards put forward for balancing the GDPR forgetting requirements, should need strong support from both the business and academic community in order to be widely adopted.

10.1.2 Big Data Analytics

As we have demonstrated in this book, big data profiling and automated decision practices, albeit powerful and pioneering, they are also highly unregulated and thereby unfair and intrusive. In fact, their regulation is currently in infant stages, allowing thus the vast amount of personal data collected by public and private bodies to make people more transparent to authorities and corporations without applying the principle of transparency vice versa, to make authorities and corporations more transparent to citizens and consumers [8]. Therefore, it is argued that the principles of accountability, transparency, and interpretability need to be clearly and unambiguously addressed in any big data analytics regulatory framework. With this in mind, the GDPR attempts to govern algorithmic decision-making and profiling by introducing transparency as a basic element of algorithmic accountability [9]. In addition, the Guidelines published by the Council of Europe "*on the protection of individuals with regard to the processing of personal data in a world of Big Data*" [10] constitute another attempt for regulating the harms arising from the big data analytics practices. However, both legal documents are not enough on their own to protect against the threats arising from these advanced technologies because, on the one hand, they are not exhaustive and, on the other, they cannot be enforced globally. Still, these same limitations offer opportunities to data scientists and legal experts for elaborating more on these practices and for cooperating towards transforming legal requirements to technical standards which could be easily applied to all algorithmically automated decision systems.

10.1.3 Backups and Archives

As our analysis demonstrates, tampering with backups, regardless of whether it is intended or not, is neither a trivial task nor a straightforward process and it is heavily impacted by factors such as the data retention regulations and the mediums used for backups. Clearly, applying the RtbF requirements on organizations' long-term archival storage it may not only severely affect business operations on tracking and discovering personal information within backed up and archived data, but it will also impose major challenges on advanced ERP data analytics and automated business decisions. Above all, there will be profound implications both for the relevant standards and for the search and indexing services.

First and foremost, the security standards stipulating backup processes need to be amended in order to be aligned with the GDPR provisions. This is necessary in order for the standards to cater for cases of data erasure requests and consequently to safeguard the effective implementation of GDPR-compliant backup and archiving search services. Technologically speaking, the thus far available technology seems to fall behind in methods for efficient search algorithms capable of looking across the entire data landscape in a cross-platform and a cross-format manner without any noticeable delays. But while these challenges do seem discouraging, we firmly believe that they offer at the same time great opportunities for expanding the current technological limits of data processing. For instance, a potential promising direction for further research is the one that pursuits the evolution of clever index and search algorithms which can extract knowledge from a disparate set of structured and unstructured data being both at rest and in transit. At the same time, improving backup mediums to allow high quality long storage with unlimited number of rewrites is another field that needs to be additionally explored.

10.1.4 Blockchain

Considering that the GDPR has not taken into account emerging decentralized p2p technologies such as blockchains—let alone the IPFS—when legislated the RtbF, its impact on such networks is rather severe. As we have shown in this book, the RtbF conflicts with the immutable nature of the blockchain. Notwithstanding the considerable research carried out nowadays to design and develop methods for conditional editing public blockchains while maintaining their security, auditability, and transparency, the conflict between the RtbF and the blockchain—even though may be temporarily mitigated by adopting off-chain workarounds like IPFS—persists. Taking further into account that this limitation may substantially affect the adoption of public permissionless blockchains to a broad area of applications, we believe that resolving such disputed areas can benefit both industry and academia.

While, as shown in this book, there are already a number of research works based on cryptography that allow the erasure of data held in private blockchains, this does not apply for the case of public permissionless blockchains whose editing and redaction is still immature. On account of this, further research efforts need to concentrate on exploring advanced cryptographic primitives to improve the trust assumptions and the consensus mechanisms in public permissionless blockchains in order their secure and accountable editing to be possible.

10.1.5 *IPFS and Other Decentralized P2p File Storage Systems*

IPFS is used widely nowadays to distribute data efficiently across numerous peers without the limitations of location-based addressing. In combination with blockchains, it offers a flexible solution to privacy issues arising from the conflicts between the GDPR and the blockchains. Yet, the IPFS—like most decentralized p2p file storage and sharing networks—suffers also from the same incompatibility with the RtbF [11]. To that end, in this book we introduced an anonymous protocol that can disseminate a delegated content erasure request securely across the IPFS network in order to resolve the identified gap between the IPFS and the RtbF elegantly.

In the future we plan to investigate the suitability of our proposed protocol in other decentralized file storage systems since we believe that their compliance with the GDPR will encourage their adoption. Among others, we intent to look into the development of mechanisms to prevent limitations such as those discussed in Sect. 8.6.6. Another very interesting direction for future research is the prevention and mitigation of malware campaigns and sophisticated attacks such as those analysed in [11, 12] where malicious content has been disseminated across the IPFS nodes by exploiting the immutable nature of the network. In addition, we indent to study thoroughly the cases where illegal or copyrighted data are spread throughout the network due to the current infeasibility of detecting their new altered hashes resulted from just a slight change in their content.

References

1. K. O'Hara, N. Shadbolt, *The Spy in the Coffee Machine: The End of Privacy as we Know it* (Oneworld Publications, 2014)
2. L. Einav, J. Levin, The data revolution and economic analysis. Innov. Policy Econ. **14**(1), 1–24 (2014)
3. P. Ohm, Broken promises of privacy: responding to the surprising failure of anonymization. UCLA l Rev **57**, 1701 (2009)
4. N.M. Richards, J.H. King, Three paradoxes of big data. Stanford Law Rev. Online **66**, 41 (2013)
5. D. Lindsay The "right to be forgotten" is not censorship (2012). http://www.monash.edu/news/opinions/the-right-to-be-forgotten-is-not-censorship
6. S. Kulk, F.J. Zuiderveen Borgesius, Privacy, freedom of expression, and the right to be forgotten in Europe, in *Polonetsky J.* ed. by O. Tene, E. Selinger (Cambridge University Press, Cambridge Handbook of Consumer Privacy, 2017)
7. A. Mantelero, The EU proposal for a general data protection regulation and the roots of the "right to be forgotten". Comput. Law Secur. Rev. **29**(3), 229–235 (2013)
8. J.C. Sharman, Privacy as roguery: personal financial information in an age of transparency. Public Admin. **87**(4), 717–731 (2009)
9. M.E. Kaminski, The right to explanation, explained. U of Colorado Law Legal Studies Research Paper No 18–24. Berkeley Technol. Law J. **34** (2018)
10. Council of Europe, Guidelines on the protection of individuals with regard to the processing of personal data in a world of Big Data (2017). https://rm.coe.int/16806ebe7a

11. F. Casino, E. Politou, E. Alepis, C. Patsakis, Immutability and decentralized storage: an analysis of emerging threats. IEEE Access **8**, 4737–4744 (2020)
12. C. Patsakis, F. Casino, Hydras and ipfs: a decentralised playground for malware. Int. J. Inf. Secur. (2019)

Chapter 11
Conclusions

Abstract Privacy and data protection rights have been characterized as some of the most critical challenges of our technologically modern distributed era, where sheer amount of big personal information is exploited by the tremendous progress in technologies like ubiquitous computing and decentralized p2p networks. To this respect, in this book we identified the apparent conflicts and the potential gaps between state-of-the-art technologies and the GDPR in the distributed era, we thoroughly studied them and we provided—when possible—specialized solutions towards resolving them. In this chapter, we summarize our key findings and contributions throughout the course of our research and we conclude our book by highlighting potential pitfalls and opportunities for the rights to privacy and to data protection in our distributed era.

11.1 Introduction

Privacy is undeniably one of the most critical challenges of our modern times in which the tremendous progress of technologies like ubiquitous computing and decentralized p2p networks struggle to be counterbalanced, among others, with the free and unobstructed flow of information, the market dynamics of personalized services, as well as ethical principles such as the right to privacy and to data protection. As a matter of fact, as O'Hara and Shadbolt quote [1], "*the point about privacy is that it raises hard cases*". To the extent that the balance between the benefits of using personal data and the individual privacy rights are under continuous debate and serious doubt by legal, science, political, social and other disciplines [2–4], the problem of privacy in the era of big data and decentralized technologies will sustain.

Throughout the course of our research for this book, we identified the apparent conflicts and the potential gaps between state-of-the-art technologies and the GDPR in order, on the one hand, to thoroughly study them, and on the other, to provide—when possible—specialized solutions towards resolving them. In this perspective, below we summarize our key findings and contributions which are consisted of:

E. Politou et al., *Privacy and Data Protection Challenges in the Distributed Era*, Learning and Analytics in Intelligent Systems 26,
https://doi.org/10.1007/978-3-030-85443-0_11

- a thorough investigation about the notions of forgetting both in the technical and in the social context, as well as an analysis of the RtbF enshrined in the GDPR,
- a review on current architectures, frameworks and technologies for the effective implementation of the RtbF,
- an impact analysis of modern organizational practices such as backup and archiving procedures towards aligning them with the RtbF,
- an analysis of the GDPR provisions regarding its protection against the risks of aggressive profiling practices and discriminatory automated decisions,
- an analysis on the privacy risks of state-of-the-art technologies such as mobile ubiquitous computing and decentralized p2p networks,
- an original in-depth review on the highly controversial topic of blockchain immutability with respect to the RtbF,
- a formal proposal of a secure and efficient protocol for integrating delegated content erasure into the IPFS in order to align it with the RtbF requirements.
- an inspection of the privacy issues arising from the COVID-19 pandemic during which digital technologies are being extensively utilized to monitor people's lives.

In what follows we describe comprehensively our aforementioned contributions. Forgetting personal data and revoking consent under the GDPR: Challenges and Proposed Solutions

Following a detailed explanation on the notions of privacy and data protection in Chap. 2, our preliminary investigation concentrated on the RtbF enshrined in the GDPR. As the RtbF is considered a highly radical and controversial right due to its drastic consequences when enforced in the era of big data, blockchains, and the IoT, we elaborated on the various notions of forgetting and the need to be forgotten both in the social and in the technical context (Sect. 3.5.1). Provided that the case of consent revocation can potentially trigger the effective erasure of personal data under the RtbF, the *right to withdraw consent* was also analysed and its various misuses were discussed (Sect. 3.4). Furthermore, we disentangled a common misconception about the GDPR RtbF and the one enforced by the CJEU decision at the Google Spain case in 2014 (Sect. 3.5.2). Then, we evaluated the implementation challenges of enforcing the RtbF in the digital environments (Sect. 4.2).

Our main contribution on this subject, however, derives from our research towards pursuing the effective implementation of the GDPR's requirements for revoking consent (Sect. 3.4.3) and permanently deleting widely disseminated personal data under the RtbF (Sect. 4.4). In that respect, we evaluated existing methods, architectures and frameworks, existing either in business or academic environments, in terms of fulfilling the technical practicalities for the effective integration of the RtbF requirements into current computing infrastructures. In particular, we highlighted their weaknesses and strengths in reference to users' full control over their personal data and their effective erasure from third party controllers to whom the data have been disseminated. While the list of the methods referenced and evaluated is not be exhaustive, they signify undoubtedly the potentiality and feasibility of implementing forgetting in modern IT systems and in digital ecosystems. Whether these approaches will apply successfully into our digital world, it remains to be seen.

Backups and the RtbF in the GDPR

Beyond other challenges that organizations have to face following the GDPR's enforcement, erasure requests from backups and archives have also become a thorny issue. Against this background, we delved into the impact of enforcing the GDPR on already established backup and archiving procedures stipulated by the modern security standards (Sect. 4.3). In our initial analysis we identified the GDPR loophole regarding the exemption from the "forgetting" requirement for archiving and research purposes (Sect. 4.3.1), whereas we recognized the most prevailing international security standards that specify backup procedures (Sect. 4.3.3). We then examined the implications of erasure requests on current IT backup systems and we highlighted a number of envisaged organizational, business and technical challenges pertained to the widely known backup standards, data retention policies, backup mediums, search services and ERP systems (Sect. 4.3.4).

As our analysis demonstrates, tampering with backups, regardless of whether it is intended or not, is neither a trivial task nor a straightforward process and it is heavily impacted by the data retention regulations and the mediums used for backups. Hence, applying the RtbF requirements on organizations' long-term archival storage it may not only severely affect business operations on tracking and discovering personal information within backed up and archived data, but it will also impose major challenges on advanced ERP data analytics and automated business decisions. Above all, there will be profound implications both for the backup standards, which need to be inevitably aligned with the GDPR provisions, and for the search and indexing services, which should expand the current technological limits of data processing.

Big data privacy challenges and the GDPR

As privacy and big data are in many ways contradictory to each other, the strict provisions of the GDPR on personal data processing is seen as being incompatible with the progress of big data analytics, machine learning algorithms and ubiquitous computing. On that account, following our introduction in Chap. 5 to the two state-of-the-art technologies of our times, mobile ubiquitous computing and decentralized p2p networks, we analysed and discussed the risks to privacy imposed by ubiquitous computing practice and research when combined with big data algorithmic processing, and we highlighted their implications for individuals on the basis of the GDPR and the RtbF (Sects. 6.2.1–6.2.3). We specifically delved into the risks of profiling which are further elaborated in the tax and financial context (Sect. 6.2.4). To this respect, we considered the recent international and European policies towards financial and tax transparency, along with their implications for privacy when used for profiling and automated decision purposes (Sect. 6.2.5). We demonstrated that even though most EU policies aim to promote effective political and legal responses for enabling an innovative, transparent and with equal opportunities economic environment, most of the times they disregard data protection values and they severely affect people's privacy, especially when these policies are combined with big data technology and algorithmic processing for profiling citizens and consumers. In light of this, the enforcement of the GDPR compelled many thus far established, or even recently enforced, international policies and legislations on tax and financial sector to be re-evaluated regarding their compatibility with its data protection principles.

Beyond the above, we also reviewed the academic discussions towards accountable and transparent profiling and automated decision processing (Sect. 6.2.6) and we investigated in depth the extent to which the GDPR provisions establish a protection regime for individuals against the risks of aggressive profiling practices and discriminatory automated decisions (Sect. 6.3.1). Finally, our contribution on this subject concluded with a set of countermeasures for mitigating the aforementioned risks in the context of financial privacy (Sect. 6.3.2).

Blockchain Immutability and the RtbF

Taking into account that blockchain technology emerged over the period under which the final GDPR text was being debated and finalized, its compatibility with the GDPR is challenging, if not impracticable. In view of the inconsistencies found between the immutability of the blockchain and the erasing requirements imposed by the GDPR, the RtbF's integration into the design of new technological advancements such as blockchains is currently disputable.

In this book, we first presented and examined the main characteristics and properties of the blockchain (Sect. 5.3.1) and its limitations in terms of privacy (Sect. 7.2). We then analysed the contradiction between the blockchain's immutability and the GDPR's erasing obligations both in technical and legal terms given that, on the one hand, the immutability is considered a cornerstone of blockchain's security, while on the other, non-compliance with the RtbF requirement may impose high sanctions (Sect. 7.3). Subsequently, and towards resolving this conflict, we commenced a comprehensive review on the state-of-the-art research approaches that have been introduced for balancing the blockchain's immutable nature with the RtbF, while preserving at the same time its security (Sect. 7.4). These approaches comprise technical methods and workarounds broadly used nowadays to bypass the blockchain's collision with the RtbF, as well as cryptographic and other advanced techniques aiming at the conditional editing of the blockchain. In our study, along with our detailed presentation of these technical methods and cryptographic techniques, we also evaluated them in terms of their potentials, constraints and limitations when applied in the wild to either permissioned or permissionless blockchain. In addition, we discussed the tension around blockchain's evolution and the challenges imposed by the RtbF (Sect. 7.5).

Delegated content erasure in the IPFS

Although, as we elaborated in Chap. 7, significant research is being carried out nowadays for implementing restricted mutability on blockchain environments, allowing data stored in public blockchains to be edited or deleted is still a controversial topic. To overcome this barrier while maintaining the benefits of decentralization, several blockchain projects are adopting the IPFS network to store the actual personal data off-chain in order to comply with the GDPR's RtF requirement. Yet, storing the actual files in the IPFS network does not remove the burden of erasing them should the RtbF be raised. Furthermore, the complete removal of content across the entire IPFS network has not been foreseen thus far by the IPFS protocol (Sects. 8.3–8.4).

Given the adverse implications of non-complying with the privacy and data protection rights, in this chapter we first discussed and analysed how the IPFS could be possibly comply with the RtbF since ensuring content erasure in a decentralized network is, in fact, questionable (Sect. 8.5). Our main contribution, however, derives from our formal introduction of an anonymous delegation protocol that could be adopted by the IPFS to securely distribute a content erasure request among all IPFS peers when a request for erasure under the RtbF needs to be fulfilled (Sect. 8.6). The proposed protocol is based on the current IPFS architecture, and as such it can be easily integrated into the IPFS. It also complies with the primary principle of the IPFS to prevent censoring; therefore, erasure is only allowed to the original content provider or her delegates. To support the adequacy and the usefulness of our proposed protocol, we provide a formal definition along with the security proofs and a set of experiments that prove its efficacy. Above all, we extensively discuss the benefits and the performance of the proposed protocol, showcasing its strength and its potentials.

To the best of our knowledge, this is the first application-agnostic proposal to align the IPFS with the RtbF and to endorse its GDPR compliance. Hence, we firmly believe that our innovative work adds real value to the IPFS in terms of its privacy enhancement and, consequently, contributes significantly to its future adoption by applications that are processing personal data.

Privacy in the COVID-19 Era

In the final chapter we investigated the interplay between privacy and COVID-19 pandemic and we explored digital technologies, such as the mobile contact tracing apps and health immunity passports, that have been utilized to monitor people's lives. To this end, we first introduced the basic building blocks and technical characteristics of the contact tracing mobile applications and health immunity passports, and then we discussed their impact on data protection and individuals' privacy. As the conflicts between public health interests and data protection principles are more relevant today than any other time, we highlighted the complexity of having to respect privacy principles while ensuring that public health measures are appropriately implemented to protect the citizens' health.

References

1. K. O'Hara, N. Shadbolt, *The Spy in the Coffee Machine: The End of Privacy as we Know It* (Oneworld Publications, 2014)
2. G. King, Ensuring the data-rich future of the social sciences. Science **331**(6018), 719–721 (2011)
3. M. Musolesi, Big mobile data mining: good or evil? IEEE Internet Comput. **18**(1), 78–81 (2014)
4. O. Tene, J. Polonetsky, Privacy in the age of big data: a time for big decisions. Stanford Law Rev. Online **64**, 63 (2012)

Lightning Source UK Ltd.
Milton Keynes UK
UKHW021851281022
411282UK00001B/9